The Language of Business

A TAPPI PRESS Anthology of Published Papers
1980 - 1991

by Cheryl Reimold

TAPPI
Atlanta, Georgia

Disclaimer

The papers in this publication are printed as they were presented or published, without alteration. Any advertisements may be out of date, and have been included to facilitate publication. The Association assumes no liability or responsibility in connection with the use of this information or data, including but not limited to, any liability or responsibility under patent, copyright, or trade secret laws.

The use of this material does not imply these manufacturers are the only or best sources of the equipment or information, or that TAPPI endorses them in any way.

1st Printing

TAPPI PRESS
Technology Park/Atlanta
P.O. Box 105113
Atlanta, GA 30348-5113, U.S.A.

TAPPI Keywords: Business communication; Business writing; Editing.

ISBN: 0-89852-266-8 • TAPPI PRESS #R-200
Printed in the United States of America

Library of Congress Cataloging-in-Publication Data

Reimold, Cheryl.
The language of business : a TAPPI PRESS anthology of published papers, 1980-1991 / by Cheryl Reimold.
p. cm.
ISBN 0-89852-266-8
1. Business writing. I. Title.
HF5718.3.R45 1992 92-3287
808'.06665--dc20 CIP

TAPPI's Antitrust Policy Statement

TAPPI is a professional and scientific association organized to further the application of science, engineering, and technology in the pulp and paper, packaging and converting, and allied industries. Its aim is to promote research and education, and to arrange for the collection, dissemination and interchange of technical concepts and information in fields of interest to its members. TAPPI is not intended to, and may not, play any role in the competitive decisions of its members or their employers, or in any way restrict competition among companies.

Through its seminars, short courses, technical conferences, and other activities, TAPPI brings together representatives of competitors in the pulp and paper industry. Although the subject matter of TAPPI activities is normally technical in nature, and although the purpose of these activities is principally educational and there is no intent to restrain competition in any manner, nevertheless the Board of Directors recognizes the possibility that the Association and its activities could be seen by some as an opportunity for anticompetitive conduct. For this reason, the Board has taken the opportunity, through this statement of policy, to make clear its unequivocal support for the policy of competition served by the antitrust laws and its uncompromising intent to comply strictly in all respects with those laws.

In addition to the Association's firm commitment to the principle of competition served by the antitrust laws, the penalties which may be imposed upon both the Association and its individual and corporate members involved in any violation of the antitrust laws are so severe that good business judgment demands that every effort be made to avoid any such violation. Certain violations of the Sherman Act, such as price-fixing, are felony crimes for which individuals may be imprisoned for up to three (3) years or fined up to $100,000, or both, and corporations can be fined up to $1 million for each offense. In addition, treble damage claims by private parties (including class actions) for antitrust violations are extremely expensive to litigate and can result in judgments of a magnitude which could destroy the Association and seriously affect the financial interests of its members.

It shall be the responsibility of every member of TAPPI to be guided by TAPPI's policy of strict compliance with the antitrust laws in all TAPPI activities. It shall be the special responsibility of committee chairmen, Association officers, and officers of Local Sections to ensure that this policy is known and adhered to in the course of activities pursued under their leadership.

To assist the TAPPI staff and all its officers, directors, committee chairmen, and Local Section officers in recognizing situations which may raise the appearance of an antitrust problem, the Board will as a matter of policy furnish to each of such persons the Association's General Rules of Antitrust Compliance. The Association will also make available general legal advice when questions arise as to the manner in which the antitrust laws may apply to the activities of TAPPI or any committee or Section thereof.

Antitrust compliance is the responsibility of every TAPPI member. Any violation of the TAPPI General Rules of antitrust compliance or this general policy will result in immediate suspension from membership in the Association and immediate removal from any Association office held by a member violating this policy.

General Rules of Antitrust Compliance

The following rules are applicable to all TAPPI activities and must be observed in all situations and under all circumstances without exception or qualification other than those noted below:

1. Neither TAPPI nor any committee, Section or activity of TAPPI shall be used for the purpose of bringing about or attempting to bring about any understanding or agreement, written or oral, formal or informal, express or implied, among competitors with regard to prices, terms or conditions of sale, distribution, volume of production, territories or customers.

2. No TAPPI activity or communication shall include discussion for any purpose or in any fashion of prices or pricing methods, production quotas or other limitations on either the timing or volume of production or sale, or allocation of territories or customers.

3. No TAPPI committee or Section shall undertake any activity which involves exchange or collection and dissemination among competitors of any information regarding prices or pricing methods.

4. No TAPPI committee or group should undertake the collection of individual firm cost data, or the dissemination of any compilation of such data, without prior approval of legal counsel provided by the Association.

5. No TAPPI activity should involve any discussion of costs, or any exchange of cost information, for the purpose or with the probable effect of:

   a. increasing, maintaining or stabilizing prices; or,

   b. reducing competition in the marketplace with respect to the range or quality of products or services offered.

6. No discussion of costs should be undertaken in connection with any TAPPI activity for the purpose or with the probable effect of promoting agreement among competing firms with respect to their selection of products for purchase, their choice of suppliers, or the prices they will pay for supplies.

7. Scientific papers published by TAPPI or presented in connection with TAPPI programs may refer to costs, provided such references are not accompanied by any suggestion, express or implied, to the effect that prices should be adjusted or maintained in order to reflect such costs. All papers containing cost information must be reviewed by the TAPPI legal counsel for possible antitrust implications prior to publication or presentation.

8. Authors of conference papers shall be informed of TAPPI's antitrust policy and the need to comply therewith in the preparation and presentation of their papers.

9. No TAPPI activity or communication shall include any discussion which might be construed as an attempt to prevent any person or business entity from gaining access to any market or customer for goods or services, or to prevent any business entity from obtaining a supply of goods or otherwise purchasing goods or services freely in the market.

10. No person shall be unreasonably excluded from participation in any TAPPI activity, committee or Section where such exclusion may impair such person's ability to compete effectively in the pulp and paper industry.

11. Neither TAPPI nor any committee or Section thereof shall make any effort to bring about the standardization of any product for the purpose or with the effect of preventing the manufacture or sale of any product not conforming to a specified standard.

12. No TAPPI activity or communication shall include any discussion which might be construed as an agreement or understanding to refrain from purchasing any raw material, equipment, services or other supplies from any supplier.

13. Committee chairmen shall prepare meeting agendas in advance and forward the agendas to TAPPI headquarters for review prior to their meetings. Minutes of such meetings shall not be distributed until they are reviewed for antitrust implications by TAPPI headquarters staff.

14. All members are expected to comply with these guidelines and TAPPI's antitrust policy in informal discussions at the site of a TAPPI meeting, but beyond the control of its chairman, as well as in formal TAPPI activities.

15. Any company which believes that it may be or has been unfairly placed at a competitive disadvantage as a result of a TAPPI activity should so notify the TAPPI member responsible for the activity, who in turn should immediately notify TAPPI headquarters. If its complaint is not resolved by the responsible TAPPI member, the company should so notify TAPPI headquarters directly. TAPPI headquarters and appropriate Section, division, or committee officers or chairpersons will then review and attempt to resolve the complaint. In time-critical situations, the company may contact TAPPI headquarters directly.

Statement of TAPPI Antitrust policy regarding submission of copies of correspondence to TAPPI headquarters

TAPPI headquarters needs to remain aware of what particular committees and sections of TAPPI are doing or planning to do in order to better assist those groups in achieving their objectives and to continue to supervise actively the antitrust compliance of TAPPI. The Board of Directors of TAPPI therefore has adopted this formal statement of TAPPI's policy which requires that persons corresponding or receiving correspondence on behalf of TAPPI provide copies of the type of correspondence outlined below to the appropriate liaison person at TAPPI headquarters.

For this policy TAPPI does not require copies of routine, written communications regarding arrangements for speakers, meetings, travel, dinner reservations and the like.

TAPPI headquarters does require that copies of correspondence of an important nature and of non-routine matters be supplied in a timely fashion to TAPPI headquarters personnel connected with the committee or section involved as shown below:

1. Plans regarding the activities of TAPPI committees or sections.
2. Communications with other TAPPI committees or sections.
3. Communications with persons or organizations outside TAPPI.
4. All written or recurring verbal complaints or criticisms of TAPPI activities.

All correspondence falling under the above-stated policy must be forwarded promptly to the appropriate TAPPI headquarters liaison person, preferably at the time of transmittal or receipt.

The Language Of Business
A TAPPI PRESS Anthology, 1980 - 1991

Table of Contents

Chapter 3
The Tools

Chapter 4
The Types

Better Speaking and Listening

Chapter 5
The Secrets of Communication

Chapter 6
The Daily Dialogue

Chapter 7
The Group Encounter

Deanne Durney

Billy's Yellow Blanket is a picture book based on a little boy's experiences with his security blanket. The story is focused on how the child uses his imagination with his blanket in his daily life. This focus differs from other picture books about security blankets. Other books focus on children having to give up their blankets as part of growing up.

The security blanket can be a significant part of a child's life. There is an organization called Project Linus that focuses on this significance. Project Linus is a volunteer organization in which people make and then donate blankets to children who are seriously ill or injured. These blankets are given to the children to comfort them.

The buyer of *Billy's Yellow Blanket* would most likely be a parent of a child with a security blanket. The parent will buy the book because it is based on a topic that the child is familiar with. The book also shows how a child can use imagination when there are no playmates around. The book could be marketed in bookstores, toy stores, and hospital gift shops. The book could also be sold in a package that includes a blanket.

The publisher would benefit because there are few picture books about children and security blankets. Another benefit would be the different focus of the story compared to the stories that are now on market.

Foreword

For over ten years the best read page of *Tappi Journal* (outside of the cover) has been the last editorial page in the magazine—Cheryl Reimold's column "The Language of Business." This column has provided more guidance for readers who need to write, than any book or course ever could.

In her column, Cheryl has taught *Tappi Journal* readers how to master the most fundamental form of communication, the written word, in a business-world context. She has not strayed from her mission of teaching those who know something important how to get the message to those who need to know important things.

Her understanding of the business world led her to write columns which taught subordinates how to communicate upwards through the "food chain" of a corporation and, in turn, offered guidance for managers on how to effectively communicate with those employees they rarely see, yet are key to their success. This understanding has made her columns valuable tools for scientists, engineers, programmers, managers, superintendents, and others who need to use writing as a part of their job.

After ten years of columns in the Journal, it became obvious to many of our members and readers in the paper industry that the earlier columns were often lost or not easily accessible. New members and readers were frustrated in their attempts to find copies of some of the columns, and for those of us who saved them, our own copies were getting dog-eared and worn. To ease the burden of the search for those older columns, and simply to provide a handy reference, TAPPI PRESS has gathered the best of Cheryl's columns together into a book.

These columns have been arranged by Cheryl into a useful reference by using the topics most commonly needed by a fledgling writer as the chapter headings. These include: How to Write Anything Well; The Words: How to Choose Them, Use Them, or Put Them Together; How to Write Excellent Memos, Letters, and Reports; and more. In addition to her emphasis on writing, Cheryl also covers the basics of verbal communications, how to run meetings, how to make presentations, and how to use communication as a quality improvement tool.

I do not think there is a better reference work on business writing for people in the industry than this one. The reader, no matter whether novice or master of the written and spoken word, will delight in reading this book. It comes in small, easily read sections (discrete bits, for those of you with a math bent) that can help anyone improve a weak area, and never fail to add even more to a strong one. This is the kind of book that begs to be used by the practical writer and communicator, and pays dividends beyond the expected.

Matthew Coleman
Manager, Communications
TAPPI

Introduction

This book will show you how to communicate well in any situation in business.

There. You have just had your first lesson in effective communication: **Give your reader your main message first.**

That's what I did here. I thought, What is my main message to the readers of this introduction? What do I want them to know—in one sentence?

Well, I wanted to tell you what this collection of *Language of Business* columns can do for you. So, instead of starting with a description and history of the columns or a treatise on the value of communication (both of which I was tempted to do!), I made myself give you the main message first: This book will show you how to communicate well in any situation in business.

Putting your main message first is a vital principle of reader-focused writing—writing that aims to meet the reader's needs. This is the type of writing that works in business, because it tells busy people what they want or need to know quickly, clearly, and completely. If you start with your main message and go on to organize your piece to meet your readers' needs, even the busiest people will read what you've written. The proof? You are a busy person, and you've just read over 200 words of this introduction!

A collection of ten years of columns published in *Tappi Journal,* this book gives you short, practical guidelines and examples for just about every type of communication in business—particularly the pulp and paper business. Most of the examples come from memos or reports written on papermaking.

In the first part, you will learn how to write effective, reader-focused memos, letters, and reports. In the second part, you will see how to speak and listen well. And in the last part, you will discover other ways in which communication can be one of your most powerful quality-improvement tools.

The first five chapters present selected columns on effective writing.

Chapter 1, **The Basics,** shows you how to approach any writing task. The columns in this chapter focus on the *thinking* that precedes and parallels the writing. You'll see what good business writing is—and what it emphatically isn't. Then you'll learn how to separate the distinct tasks of writing and editing, adopt the right tone for a piece, present technical material effectively, and organize your writing to transmit your ideas accurately and persuasively.

Chapter 2, **The Words,** guides you to the right choice and usage of words and gives you techniques for putting them together into powerful sentences and paragraphs.

Chapter 3, **The Tools,** answers your questions on grammar, punctuation, and spelling and teaches you how to write helpful headings and create effective illustrations.

Chapter 4, **The Types,** shows you how to apply all the ideas and techniques of the previous chapters to each category of business writing, with specific guidelines for various types of letters, memos, and reports.

The next three chapters cover effective oral communication—speaking, listening, holding meetings, and giving presentations.

Chapter 5, **The Secrets of Communication,** reveals some vital keys to effective communication. You will learn some special lessons from great communicators, discover the masks we often wear and see how to remove them, and find six ways to break through barriers to get your intended message across.

Chapter 6, **The Daily Dialogue,** takes up the various instances of communicating on the job. You'll find practical techniques for getting your message across, negotiating with opponents of your ideas, and dealing with other difficult situations.

Chapter 7, **The Group Encounter,** provides guidelines for holding productive meetings and giving winning presentations. You will see how to run constructive meetings by developing a solid purpose, structure, agenda, and "meeting culture." And, through the use of practical techniques, you will learn how to conquer stage fright and give powerful, well-structured presentations *without fear!*

The final section consists of a single chapter of columns on communication and quality control.

Chapter 8, **The Key to Quality,** shows you how to derive extra, valuable benefits from writing and speaking skills. You'll discover how to use writing to solve problems and significantly improve the overall quality of your work. You'll also find practical techniques for managing writing effectively in your group, and some principles from poetry (of all things!) for sharpening your professional and managerial skills.

I hope you enjoy reading the book and mastering the language of business. Good luck!

Chapter 1
The Basics

How to Write Anything Well

The columns in this chapter offer a practical approach to anything you ever write in business.

We begin with an overview of effective business writing. The first three columns—**Business writing—clear and simple, What's your message?** and **How to stop hating writing**—show you how to define your message and break through the clutter of corporate jargon to transmit it effectively and readably.

Then come some general writing techniques. **Writing and editing,** a two-part series, explains a new way of drafting, shaping, and completing your piece. **Tone** and **The message that always comes through** show you how to transmit a positive attitude in your writing. **Please—do take it personally** tells you how to write readably. **The gentle art of persuasion** presents techniques for getting the reactions you seek. And a two-part series, **Communication ≠ Information,** gives you a new way to think about communication and shows you how you can always meet your readers' needs.

The next three columns take up technical language. **Science and language—the search for elegant simplicity** presents four guidelines for good scientific writing. **Technical writers: loosen up** shows you how to avoid typical pitfalls in technical memos and reports. And **Are you an impressionist or an expressionist?** describes what happens when you ignore the rules of clear, straightforward technical writing.

The final columns in this chapter take you straight to the pen or the word processor. **How to overcome writer's block** presents a personally tested and guaranteed method for getting started. **A universal outline** describes a plan that will work for *everything* you ever write. A three-part series,

Beginning, middle, and end, takes you through the entire organization of an effective piece of writing. And **Getting your ideas across** gives you ten ways to do just that.

Business writing—clear and simple

On April 15 of this year, when ordinary citizens were busy preparing their tax returns, New York Governor Hugh Carey was busy with language. He issued an executive order calling for state agencies to rewrite their regulations in plain, understandable English. This order stated, in part:

> When pursuant to subdivision three of section two-hundred-two of the State Administrative Procedure Act an agency submits to the Secretary of State a Notice of Action Taken in which it is stated that there are substantive changes in the final action in comparison with the proposed rule or regulation, the agency shall also summarize the changes, if any, which must be made in the regulatory impact statement in order to reflect accurately the impact of the final action taken.

In other words (74 fewer): keep impact statements (whatever they are) up to date.

Bureaucratic verbiage, you sigh. But—what about the language of business? A mill manager writes that he "anticipates the possibility of requiring an additional operative" when he wants to hire someone. A vice-president for research sends his staff a memo complaining that "there is an excessive time frame existing between your productive effort and the external communication of that effort" when he wants them to get their technical reports out faster.

Corporate verbosity—does it matter?

Think about your own reaction to pompous, wordy reports. You hunt for the message hidden in the clutter, skipping paragraphs, scanning pages. You get distracted. You get confused. You get a headache. And, eventually, you stop reading.

Clear, effective language is essential to technical work. Very often you are the only one in the company who knows all about some equipment or procedure—and you have to explain it to others. You must be able to tell the people who run the corporation just what you are doing and why they should continue to support it.

You must be able to *tell someone something,* without smothering it in puffs of polysyllabics and technical jargon.

How do you do it?

First, weed your garden. Clear out every vague or repetitive word that stifles the clear expression of your thought. When you find yourself:

- using the long word or expression instead of the short
- using technical jargon when a plain English word would do as well
- seeking to sound important, impressive, or knowledgeable
- using as many fuzzy words of as many syllables as possible to cover areas of ignorance—

then stop and try again!

Once your "garden" is clear, you can start planting good linguistic habits. Begin every piece of writing by asking yourself:

1. *What* am I trying to say? Jot it down in plain English. Then you can look back to see if you've said it.

2. *To whom?* If your readers are not all specialists in your technical field, avoid technical language as much as possible. You may protest that the technical words are the only precise expression of the thing itself. Then ask yourself what a precise expression in Greek conveys to an English-speaking reader. And say it less precisely in plain English.

3. *How much* do my readers know about this? Even nontechnical readers may be well-informed about the subject you're discussing. Don't bog them down with things they already know. Tell them only what they need to understand your article.

4. *Why* am I writing this? The purpose of the piece will largely determine its style and emphasis. Is it primarily to give the readers specific information? To persuade them to do something? To convince them? Remember, the purpose is *never* to glorify the writer! When you forget this, you'll find yourself writing convoluted sentences, dazzlingly long and obscure words, and 10 pages instead of the necessary 1 or 2.

No extra words

Now you can start writing. Hold onto one fundamental principle: every word should work. No idle words in your prose; each must justify its existency by meaning.

Write as you would talk. That means using "lever," not "manipulandum," and "cost" instead of "dollar volume."

When you've finished, make a short revision checklist:

- Did I say what I wanted to say?
- Did I fulfill my purpose?
- If this arrived on my desk as is, *would I read it through?*

We'll look at the specific components and the arch-enemies of good, lively writing in upcoming articles. Yes, it matters. A director of research of a large corporation put it succinctly:

> If you can't tell in written or oral English what your results are, it is impossible to get along in any industry. If you can't put your thoughts and figures on paper in concise readable language—you're sunk!

Ms. Reimold is a professional writer who contributes to the Macmillan Book Club and the McGraw-Hill Book Club and who has held positions with Westchester Illustrated; *Holt, Rinehart and Winston; and the public relations firm of Burson-Marsteller. This is the first in a series of articles she will present on effective communications skills for engineers and managers.*

What's your message?

Imagine biting into a jelly donut and finding no jelly. Or opening a birthday card and finding no signature. How would you feel? Disconcerted? Frustrated? Maybe even angry?

That's what happens to people who are forced to read memos and reports that have no clear message. And it happens in business all the time. In all the letters, memos, and reports I see, the greatest problem is lack of a clear main message. A technical report gives the details of a project but doesn't explain the significance of these details, or of the project itself. A memo announces a meeting but doesn't say why the reader should attend. In both cases, the writer omitted a crucial step in preparing his draft; he didn't ask himself why he was telling this reader these things.

What readers want—and what writers do

A memo is delivered to your desk. As you pick it up, what are the first thoughts that go through your mind? Don't read on for a minute. Close your eyes and think about it.

Was your answer something like: "What's this about? Is there anything of interest to me in it?" Those are the two questions busy people want answered right up front. The answers constitute the writer's main message. If you provide those answers consistently in your first paragraph, you will be a prized member of your organization for you'll be one of the few who make their message clear.

Everyone wants other people to state their message clearly, up front. But most of us are reluctant to do so ourselves. Why? I think we're afraid the reader won't read on if he knows what it's all about. Better to keep him guessing, we reason. At least that way we'll keep him reading!

Don't try to delude people into reading something you think won't interest them. Instead, think of something about your topic that will interest them, then state up front what you have to say and why they'll want to know about it.

Find your message

Your message consists of your answers to two questions; answer them before you start writing. Write out your answers in complete sentences—any other way won't work:

1. What do I want to tell them (him/her)?
2. Why would it interest them?

Your answers should appear in your first paragraph.

In my writing classes, I often do a dialogue with the students to help them find and state their message. You can do a similar dialogue in your own mind. It goes something like this:

T(eacher): Imagine I'm your reader. What do you want to tell me?

S(tudent): About the safety seminar next week.

T: A complete sentence, please, with the relevant information.

S: There will be a safety seminar in the conference room next Wednesday at 9 a.m.

T: Right. But the safety seminar is the subject of your memo. What do you want to tell me about it?

S: Well, I want you to come to it.

T: Why should I? What would interest me there?

S: You'd learn how to avoid serious accidents in the lab. We're going to explain what can happen and show people the safety equipment they should wear when entering certain rooms.

T: Good. So, what's your message? Remember, it includes what you want me to know and why it'll be of interest to me.

S: Okay, here goes. "Come to the safety seminar in the conference room next Wednesday at 9 a.m. You'll learn what safety equipment you need to wear to avoid serious danger in some rooms in this lab."

T: There's your first paragraph. Great. Here's another example, for a memo reporting test results. What do you want to tell me, your reader, in a whole sentence?

S: I have the results of the tests you wanted us to run.

T: Okay. Anything of particular interest to me in these results?

S: Let's see. Yes, they show that the uneven caliper on Sample B is probably a result of an improperly maintained swimming roll in the press section.

T: Fine. How will you state your message, then?

S: "Here are the results of the tests you asked us to run. They suggest that the uneven caliper on Sample B is probably a result of an improperly maintained swimming roll in the press section."

T: Great. As your reader, I'd appreciate knowing that right away, before I get to the tables.

Before you write, ask: What's my message? To get it, answer these two reader questions: what do you want to tell me; why would it interest me? The answers are your message. Put it in paragraph one. Your readers will love you for it.

How to stop hating writing

The following dialogue, with small variations, occurs at least once at the beginning of every writing course I teach:

C. R.: This sentence is not clear.

Student *(wringing hands or expressing pain in some other way)*: I know I don't know how to say it. I spent hours over it.

C. R.: What are you trying to say here?

Student: Well, I'm trying to explain that . . . . (He/she then explains it. For example: ". . . that the lab didn't give us enough samples for us to make a conclusive study. We'd need at least 50 to feel secure about our findings, but we had only 13.")

C. R.: Then write that.

Student: What? Write what?

C. R.: "These conclusions are tentative. To feel secure about our findings, we'd need to analyze at least 50 samples, but for this study we had only 13."

Student *(who had written something like this: "Due to a lack of adequate and sufficient data available from which to perform the analysis, the conclusions of the present study should not be considered as definitive but should be held in abeyance pending the future transmission of an additional set of samples, preferably in a quantity of at least three times the present amount.")*: Just that!

C. R.: Just that.

Student *(pen poised)*: Er—could you say it again?

C. R.: No. You say it again. Then write down what you've said. You can line-tune it later.

Why could the student say clearly what he couldn't write clearly? One major reason is a fundamental difference between writing and speaking.

That difference is the other person. When you talk, the listener is right there in front of you or on the other end of the phone line. You can see or hear his reactions; you can't avoid them. And he can see and hear you.

But when you write, there's no one reacting to you. Since you can't check how you're being received, you run the risk of being misunderstood and misjudged. Fear of being misjudged often causes the business writer to do two silly things:

1. Try to obliterate his personality from his writing altogether, reasoning that if he can't be seen, he can't be judged. Hoping to sound like every other business writer, he writes in a stilted, formal style using long words and complicated sentence structures.
2. State things in as fuzzy and uncommitted a way as possible, again hoping to avoid wrong judgments or reactions by giving the reader nothing to which he can react.

The result is the kind of writing we all hate to read: clear thoughts translated into wishy-washy, unclear language; long, convoluted sentences; jargon; pompous-sounding polysyllabic puffery; and hazy references to things happening instead of clear statements of who did what and what the writer thinks about it.

Because it's such an effort to turn straightforward thoughts into this kind of murky elusion, many people in business hate writing. And because the result is often an incomprehensible mess, many people hate reading the memos, letters, and reports that appear on their desks every day.

So we now have an absurd scenario. For at least 25% of his worktime (a conservative estimate, according to recent studies), a busy professional agonizes over writing pieces that other busy professionals will hate reading (or, for self-preservation, will not read at all). With all this writing-related agony sapping their strength, it's a wonder people have any energy left for work.

Is this effective communication? Will this promote high-quality teamwork? Obviously not.

Can you do anything about it? Yes.

Talk out your memos, letters, and reports

Literally read your writing out loud. See if it sounds honest and flows naturally. If it doesn't, think hard before you change a word here or there. Were you trying to cover up your thoughts or personality because of fear of misjudgment? If so, remember that writing born of that fear is always writing people hate to read.

Once you've faced your motivations and corrected them, try saying the offensive sentence another way.

Ask yourself, "What on earth am I trying to say here?" Say it. Then write it.

Here are three guidelines to keep in mind when you talk your writing:

1. Determine to say clearly, whenever possible, who did (does, will do) what.
2. Mistrust words of three syllables or more. If you've used them to sound impressive, replace them with a more easily spoken word or phrase.
3. Do not write any sentence you would not say.

Talk your writing—and you'll relieve the agony on both sides of the document.☐

Writing and editing—the two halves of language

Part 1: A new way of writing

East is East and West is West and never the twain shall meet . . .

As far as the West is from the East . . . so far should the act of Writing be from the act of Editing. They are distinct, independent activities, performed, it seems, by different halves of the brain. Writing is putting your thought into words. Editing is making that verbal expression palatable and understandable to the people who will read it.

Does this separation of tasks appear obvious? Perhaps—until we look at our own work. Pull out a first draft of something you wrote, a draft for a letter, a memo, a part of a technical report, anything at all. Do you see sentences begun, then crossed out and abandoned? Are there words written, struck through, changed—then perhaps written all over again? If so, you have *not* separated the writing and editing functions. You have, like almost everyone who picks up a pen or a pencil, begun to write and edit all at once.

Why change?

Now, what is wrong with this? Why do I suggest writing with *no* changes allowed and *then* editing? Two reasons.

First, you cannot *express your thoughts* clearly in writing if you're occupied in *correcting your writing* at the same time. In an excellent book on the subject, *Writing with Power*, Peter Elbow says that the writer should write his first draft for at least ten minutes without stopping—just to separate the producing from the revising process. At this stage, we should not be thinking about "how to write." Rather, we should be focusing on the subject of our discourse and allowing our creative energy to express our thoughts freely. You can see why. If you're half-focused on describing your latest experiments with freeze drying and half-focused on the words you're using to describe them—you will do each job half-well, at best.

The second reason for writing first, without editing, concerns the content of your work. If you write down all you know about the subject with no corrections or constraints, you will find you know and can express a lot more than you thought. It's like sending a plumb line down to the depths of your knowledge and experience and pulling back all that's there, with no interfacing signals to knock you off course.

So, when you sit down to write, just write. Do not allow yourself to cross out or change a single word. No stopping sentences midway, either. Let your thoughts on the subject flow their way. It may sound easy—until you try it. For to write purely like this is to break the habit of a lifetime. We have all been conditioned to write, cross out, and start again, hobbling along painfully to the end of our messy pages. Why? Mainly, I think, we don't want to waste time. We feel that if we can write and correct *all at once*, we'll have the job done in half the time. Deliberately saving the correcting for after the writing tugs at the time-constricted heartstrings of the busy twentieth-century scribbler.

Time's a wastin

There is only one way to overcome this fear of time-wasting. I tried it, and I have written first and edited second ever since. Time yourself. Make a strict account of every minute spent writing "the old way," from the moment you pull out the sheet and stare angrily at it to the moment you give it up for final typing. Then, try writing a similar project "the new way." Time yourself again. The new way goes like this:

Take out a sheet of paper and write your topic across the top of it. Begin to write about it. Write anything and everything that comes into your head on the subject, in the order it appears to you. *Force* yourself not to alter a single word. (The effort will send you into a spin the first time, but future writing will prove it's worth it. Over the years, you will save hundreds of hours.)

You will notice two things. First, after the first page or so, your speed of writing will pick up noticeably—because you are gradually freeing your creative faculties of critical clamps. Secondly, you'll be touching on aspects of the subject that you hadn't thought of before. Your hand will hardly be able to keep pace with your articulate thoughts. And—you will feel exhilarated, for what you have just done is allow your creative forces full, free rein.

After you have written all you want to write on the subject, stop. Check to see how much time you spent on that phase of your work. Then put the writing aside and do something else. Even if this pause is brief, do take it. Give your critical faculties a chance to approach your writing with a fresh start.

Now look at what you have written as if you were a third person examining it. You will find that you *feel* like a different person from the one who wrote the draft, for you are now approaching it wholly from the critical viewpoint. Before, your angle was wholly expressive. Now you are ready to edit your work.

Be your own guinea pig

Before we come to editing, the second half of the world of writing, I would like you to make an experiment. Set aside ten minutes, today. Select a topic that you will have to write about in a memo, a letter, or a report. Take out a pad of paper and write across the top of it. And then, for ten whole minutes, write about that topic without stopping, with *no* corrections, *no* fresh starts. At the end of the ten minutes, stop writing. Put your paper away, and try to resist looking at it, preferably until tomorrow! I think you will be surprised at what you see and the way you feel.

Keep the piece of writing until the next column reaches you. Then you'll see how to shape it through careful, systematic editing.

Writing and editing—the two halves of language

Part 2: Editing—the shaping of a manuscript

Most of us don't like to edit our work. We quake at the scourge of the red pencil that slashes through our manuscript, changing hard-sought expressions, questioning others, even amputating whole sentences or sections. If *only* our original outpouring could satisfy our colleagues, our readers, our editors—ourselves. Alas, it rarely does.

But—the new way of writing offers an alternative. You simply don't edit anymore. Rather, you *shape* the free-flown words into an orderly, attractive piece of communication. And it feels entirely different.

Shaping is a positive, creative act. It is what a sculptor does to a hunk of marble, a prize confectioner to a lump of dough. Shaping makes the critical difference between an unfathomable sea of thoughts and a well-charted channel between writer and reader. The red pencil becomes an instrument of creation, not amputation.

Find the unifying force

A piece of technical writing usually has a few points to make about a single topic—the subject of a memo, the event or plant described in a trip report, the experiments detailed in a technical report. In a *unified* piece, these points are connected by a single unifying force or impulse.

The *subject* of the Gettysburg Address is a memorial to the soldiers who died at the battle of Gettysburg. The *unifying force,* the impulse behind the words that draws them all powerfully together, is Lincoln's desire to preserve the Union. The short address covers a great deal—the birth of the United States, its early principles, the Civil War, the battlefield, the heroic death of the men who fought there, the charge now given to the living. Yet throughout, the words seem to throb with the energy of that single impulse: this nation must survive.

Our regular writing tasks seldom grow from a force so powerful and urgent as this. But we always write to say something. If we have nothing important to say, we'd better put down the pencil.

Read through your first unedited (I hope!) draft. Then close your eyes and try to formulate a one-sentence answer to the question: What do I want to convey? It may help to begin your answer with the words, "I want . . ." For example:

- I want you to continue supporting these experiments.
- I want our company to consider buying the copiers I examined at the XYZ plant.
- I want better service.
- I want you to understand exactly how this machine works.

Very often, the unifying force is discernible only after you've written the first draft. You have to write down all your thoughts and discoveries on a subject to find out or clarify what, essentially, you want to say about it. That is why it is much better to write first and shape later. You can be more confident of having something real to shape instead of fabricating something flimsy around a chosen focus that may be neither appropriate nor relevant.

Build your words around the unifying force

When you've found the unifying force in your writing, you will be able to extract the points you've made in its service. Underline every sentence that:

1. makes a relevant point (one line)
2. illustrates or explains a point already made (two lines)
3. connects one point or thought to another (three lines).

Check *every* sentence; if you find one that doesn't fulfill one of these conditions, it probably should go.

Now pick out your three major points. If there are more, see if any of them can be subsumed into one of the three. If not, omit them. More than three points with illustrations and explanations simply cannot be absorbed at one sitting. If your work contains a number of chapters or sections, apply this rule to each.

Make your writing good to read

Now that your work has its structure and form, you can give it the final aesthetic shaping. Make it good to read. For the final phase of shaping, you can use a checklist.

Trim away:

- **Clichés.** Expressions such as "last but not least," "in the final analysis," "in actual fact," "back to square one" are unoriginal phrases that have lost their precision through overuse and now serve to cover up the writer's personality—precisely because everybody uses them. Avoid them.
- **Irrelevant detail** that draws attention away from the unifying force
- **Excessive explanation of the obvious.** Watch out for sentences beginning with "That is to say. . ." or "In other words . . . ." If you've said it clearly, you don't need to repeat it in other words.
- **Unnecessary modifiers,** such as "a *loud* explosion," "a *high* peak," "an *empty* vacuum," or "the *final* conclusion."

Check for:

- **Correct subject/verb agreement:** Freeze drying of the sample and recommencement of the entire process three hours later *were* (not was) found to produce . . . .
- **Correct punctuation,** particularly commas that may change meaning: Next, we prepared the wood lying in the freezer . . . . *not* Next, we prepared the wood, lying in the freezer. (Unless you prefer to work in subzero temperatures.)
- **Variety** in words, expressions, and the structure and length of sentences.
- **Relationships.** Are your ideas connected? Could you make the relationships clearer by using conjunctions (e.g., *because, since, yet*) rather than placing one sentence after another?
- **Completeness.** Have you covered all you promised in your opening paragraphs?
- **Coherence.** Does your work have a beginning, a middle, and an end? Are they connected?

Shaping *is* fun. It gives form to your thoughts and makes them pleasantly accessible to others.

But, as half the writing process, it requires half the writing time. Are you willing to give it the time it needs? The reward is clear: You will be read. With pleasure. □

Tone

What is language, anyway? Is it just words—words—words? Clearly not, any more than a picture is just lines or music just sounds. Language is a combination of many things, but one of its most significant features is *tone*.

Think about it. You sit down one morning to compose a memo to your superior, a short talk to your staff, and a letter to a supplier. You don't just start to write. In each case, you mentally assume a new role, and the language of those three communications embodies that role. The *tone* of your writing is the expression of your assumed mental attitude, frame of mind, or relationship with your reader.

Trying to correct bad language habits without delving into the tone you adopt is like spraying the leaves of a tree that has root trouble: you can gloss over the problem for a while, but sooner or later it will reappear. Since the actual words you choose are so often a direct outgrowth of the tone you adopt, it is time to look carefully at the root of your troubles.

As in every cure, the first step is to recognize the problem. Take a moment to go through copies of recent memos, letters, or notes you've written to three different people. Imagine they were composed by a third person—and then describe that person. Was it: a friend trying to sound like a business associate? a business associate trying to sound like a friend? an angry customer? a willing subordinate? When you identify the role of the writer, you're capturing the tone, for the tone is the verbal rendition of that assumed role or relationship with the reader.

Now that you recognize the tone, look again. This time, imagine that you are the recipient of the piece. How do you like the tone the writer adopted? It is pompous? degrading? overly subservient? Above all, does this piece of writing make you sympathetic to the writer and agreeable to comply with his or her wishes?

You see, when you assume a role, and consequently a tone, you are by extension assuming a role for your readers. If you're the boss, they're automatically the underlings. If you're the know-it-all, they're the ignoramuses. Are the roles your writing assumes appropriate? Or have you, albeit unwittingly, caused offense?

I wonder if you are still staring at one of your memos, trying to put a name to the character who wrote it. A memo that's neither overbearing nor subservient. A tone that smacks of neither anger nor goodwill. A language that is vaguely pompous, certainly multisyllabic—ultimately *impersonal*. Here's an actual example from a recent business letter:

> As per our conversation on May 24, enclosed please find a copy of the document in question. It would be appreciated if copies of the patent could be made available so that a determination could be made as soon as possible as to whether the present outside efforts in this area fall under the umbrella of the said patent.

In "communications" such as this, all human interaction is dead. The tone is strictly impersonal. One robot could be dispatching the missive to another.

This is the tone adopted by most business and technical writers when they're addressing people with whom they have no clear relationship. They work painfully hard to prevent any vestige of humanity—writer's or reader's—from entering the text. Yet, in most cases, the writer wants something from the readers—at the very least, their attention! Do you pay attention when you're being treated as a nonentity? Would you give finding-the-said-patent top priority? Why should you?

Now, if the writer had admitted that he was a human being writing to another human being about something they have both talked about, he might have written:

> I'm enclosing a copy of the document we discussed yesterday. We would all be very grateful if you could get us a copy of the patent, as we're anxious to see if this actually falls under it. Thanks.

Here, the reader is being treated as a worthwhile, intelligent person whose help other people need and will appreciate. Surely this reader is more likely to respond with alacrity.

What is the role behind the tone of impersonality? Quite simply, the *Businessperson*. The Businessperson enters the office and, as it were, dons a business suit, scrupulously divesting himself or herself of any remaining individual characteristic that doesn't suggest *Business*. The Businessperson worships efficiency—and people are notably inefficient. So, in speaking or writing, the Businessperson pretends people, or individuals, don't exist.

An exaggeration? Think about it. Can you recognize a whiff of truth?

Again, recognizing the problem is the first major step to correcting it. Once you realize you're assuming the role of Businessperson, you will be quick to see its expressions in your writing. Some typical ones are:

- Impersonal phrases, such as:
  it has come to our attention that...
  this office is in receipt of...
  corrective action must be taken...

- "Heavy" expressions that ordinary people wouldn't use, such as:

referred to as	instead of *called*
in the amount of	instead of *for*
ahead of schedule	instead of *early*
owing to the fact that	instead of *because*
inasmuch as	instead of *as*

- A near-absence of personal pronouns.

From now on, take just a second to check, and possibly change, your role before you write. You will be amazed how quickly certain words and expressions disappear before they even reach the paper.

The message that always comes through

A young manager came up to me during a break in our writing class.

"I know you said we should write to people the way we would talk to them," she said, "but it's very difficult to do that in this organization. We're all used to writing in the passive voice and starting everything with 'as per your request'—certainly not the way we would talk. But it's actually easier for me to write in that style. Why should I force myself to put 'as you asked' instead of 'as per your request'? They both say the same thing."

"But they don't," I told her, and went on to explain why they said very different things. I think I convinced her; she told me recently that I had ruined "as per" for her forever! I hope I can convince you, too.

Choice of words transmits a message

The words a writer chooses actually carry their own message. They tell the reader what the writer is trying to convey about himself.

Let's look at an example. A group manager is seriously upset that meetings never start on time. Here are three possible memos he might send out to his people:

Memo 1: "It has come to my attention that on-time attendance at meetings in this department is increasingly delinquent. This situation cannot and will not be allowed to continue. Henceforth, the individual responsible for chairing a meeting is instructed to commence proceedings on the appointed hour, regardless of the status of attendance of participants. Noncompliance will be considered a grievous abdication of company duty."

Memo 2: "Folks, I need your help. I've heard that some of you are having trouble getting to meetings on time. I know it's sometimes hard to get up and leave work when you're in the middle of something important, but if too many people are late, the meetings wind up starting and ending later. Then everyone's unhappy. Please, people, try to get to your meetings on time. If you chair a meeting, make a real effort to start at the appointed hour, even if some folks aren't there yet. Thanks a million."

Memo 3: "We have a serious problem in our group: people are not getting to meetings on time. As a result, meetings take nearly twice as long as necessary, forcing people either to cram their own work into fewer hours or to stay late. Clearly, this is inefficient. It is also extremely unfair to those people who do make the effort to arrive on time. I am asking all of you to get to your meetings on time and to start meetings you chair promptly, no matter who may be absent. Please make no exceptions; I will support you fully in your efforts to get our meetings back on time. Thank you."

Readers react to the message

All three memos give the readers the same information, but they each transmit a very different message about the writer. Here are some possible reactions of readers to each message:

Memo 1: "He's putting on the managerial mask and using words that sound big, powerful, and anonymous—like the organization—to impress me or even frighten me enough to get me to do as he says. Not very effective; he's just an ordinary guy hiding behind big words."

Memo 2: "He's concerned about people being late to meetings, but he wants to keep us feeling good (especially about him). Maybe I should start coming five minutes late instead of ten. Don't want to agitate the old fellow too much."

Memo 3: "He's seriously upset about this. He means to follow through and get after the people who are late. From now on, I'd better get to meetings—and start my own—on time."

Notice that the straightforward memo (3) calls the least attention to the writer. This writer isn't trying to "sound like" anything. He chooses the words that will make his stated message clear, with no subscript about himself. As a result, his message gets the most attention, and he gets the most respect.

And there's the irony. If you try to sound powerful or witty or anything else, you will end up sounding simply scared. You'll transmit the message that you don't consider your own voice and self strong or powerful or clever enough, so you're putting on a mask and speaking the accompanying script.

You hurt yourself and muddy your message when you choose words to project a certain image. "Talk" your writing; don't try to "come across as" anything. You'll put power into your writing and get the respect and attention you deserve. □

Please—do take it personally

Your writing, that is. One of the most deadening characteristics of "business writing" is impersonality. People will go to almost any lengths to keep their own personality from showing in whatever they write for business.

Just think of the resume or *curriculum vitae.* It's probably the most personal written document anyone in business ever puts out. But can you think of a more *impersonal* piece of writing? In the classic, expensively prepared resumes, the word "I" never appears. The rigid categories into which the candidate's professional life is divided leave no room for individual traits or even interests. In fact, people pay for a "professional resume" precisely because they want a rigorously standard, impersonal account of their lives in business so far.

Does it work? How many people have you hired on the basis of such a resume? You may glance at the highlights to see whether the candidate is worth interviewing, but isn't that about it? Wouldn't you like that resume to give you just a hint of the person behind it?

When I was applying for a job with a large public relations firm in New York, I decided to put my disdain for the form resume to the ultimate test. I threw mine away! Then, I wrote a one-page description of what I thought the firm would like to know about me. I used full sentences, complete paragraphs, and the the pronoun "I."

The first paragraph told them what I had been doing for the past couple of years and why I thought it was good preparation for the job I was seeking. The second went a little farther back into my history, again stressing points that would interest this firm. The third talked about me: what I liked to do and what I felt I did well. It explained why I wanted to join the firm and why I thought they would be glad they had hired me.

The man who did hire me later admitted that this unusual document had done a lot to win me the position. He said that mine was the first resume he had ever actually read.

While this rather maverick approach may not be universally appropriate, its success in my case does underscore the value of keeping all your writing personal. The individual voice gives life to words; why in business writing do we try so hard to deaden them? Perhaps because everybody else does.

But when you write, whatever you write, you write to be read. That is the number one criterion. And draining the life out of a written communication does *not* make it more readable. On the contrary. People respond far more to one small piece that was clearly written by another person than to shards of standardized, dehumanized information.

Let me give you an example from an actual trip report. It begins:

> A trip was made to XYZ Corporation on (date). The purpose of the trip was to observe the operation of equipment utilized in the areas of removal of waste material. The focus of attention was the "Permafest," manufactured by ??????. Permafest is a screening machine utilizing a simple concept to separate large solid waste material from materials to be recycled...

The report goes on to describe the workings of the machine and concludes with a list of its many benefits. As there are no defects given, the reader is left to deduce that the writer was impressed with Permafest and thinks it might be an asset to his firm. But he never said so. He didn't let himself show through.

Now, if he had been discussing the trip, this writer would probably have begun by giving his personal reaction to Permafest. He would have gone on immediately to explain what it did so well and why he thought it would be a good investment. He could have written an interesting and fully informative trip report if he had approached it in this vein, beginning:

> I went to XYZ Corporation on ... to study the operations of the Permafest screening machine. After observing the machine run its full cycle several times, I was greatly impressed by its efficient and cost-effective separation of large solid wastes from material to be recycled.Permafest works on a simple mechanism and is relatively inexpensive and maintenance-free. For these and other reasons which I will explain here, I think Permafest would be an excellent asset to our mill.

Remember, the trip report is supposed to reflect the personal observations and experience of a trusted, qualified individual. One could write to the company for a simple description of how the machine operates.

Taking your writing personally does not mean being folksy or "cute." It does mean:

- Using personal pronouns—*I, we, they.* If you know *who* did, saw, or operated something, say so. You will transmit a clearer and more accurate image of the thing described than you could with the people missing.
- Avoiding impersonal language whenever possible. This includes: **1.** The passive voice (Say, "I went to XYZ Corp." instead of "A trip was made to XYZ Corp.). **2.** All expressions that tend to keep the actual human agent out of the picture. For instance, instead of "It should be recognized that," try "I noted that..." or "Keep in mind that.. ." Replace "The *x* is utilized for" whith "We (or they) use *x* to ..." Avoid "The affects of *y* can be seen" and say instead, "We saw ..." or "They demonstrated that ....".
- Rethinking your approach to everything you write. Bad tapes take a long time to erase—and many of us have been "programmed" to produce data sheets with no stamp of humanity anywhere on them. Imagine you are talking to your reader. Tell him first what you *want* to tell him. Then give him the details.
- Writing about people, not dehumanized objects. It is so simple and accurate to write: "Twelve people will be working on the project. These include." Yet a report I read said: "The total manpower required for the project is twelve agents broken down as follows." As we read, we try to translate the words into images for ourselves. The more clearly we can do this, the better the communication. I can see "twelve people"—but I can't visualize "manpower," and "twelve agents broken down" makes me want to stop reading right away!

Don't let the plague of depersonalization kill your writing. People like to hear from people, not machines. You do, don't you?

The gentle art of persuasion

Persuasion. It's at the core of practically every piece of business writing. Think about it. You write a letter to ask someone to do something for you—before doing anything for anyone else, you hope. Even when you write a "form letter"—say, a letter acknowledging receipt of a requested item—you are usually using a subtle form of persuasion (often so subtle that the *writer* doesn't realize it!). Here, you are anxious to persuade the reader of the high standing, the importance, or perhaps the integrity of yourself or your firm.

It's the same with reports. A "technical report," often treated as simply a written account of the monthly activities in the lab, is actually that lab's most powerful piece of public relations. Written well and persuasively, it will convince the readers in top management that this lab is worthy of attention *and* financial respect. That technical report is written not only to inform its readers of what's going on in research, but to *persuade* them of its importance to them and to the company as a whole.

Quarterly reports, annual reports, even inter-office memos are all a form of persuasion. More often than not, the persuasion is a hidden part of the "sub-text"—the message that lies between the lines. You don't write: "Dear Mr. President: The Company and you yourself will benefit beyond belief from the wonderful things we are doing here—if only you'll pay more attention to us and increase our budget"—but isn't that often the real message of your communiqués?

All right. You recognize that most of your letters, memos, and reports aim, somehow, to persuade. This recognition is the first step to successful persuasion, for you will now structure the pieces differently. Instead of doggedly reporting facts, you'll start thinking about the readers' reaction to your presentation of those facts. Instead of writing a chronological report—telling what you did, when, and what happened—you will find yourself organizing a piece on the basis of the readers' interests. What is top management likely to be more interested in—the order of steps or the results of a certain procedure? If the results are what they want to know about, tell them those first—even though they came last! Might your present work lead to a more efficient process that will be competitive with those used by the leading manufacturers in your field? However slim this possibility is, tell your readers about it on the very first page. It will intrigue them and make them read on.

Although there are no strict rules for persuasive writing, there are certain guidelines that will help you get the reactions you seek.

- Remember that to persuade people, you've got to draw their attention first. This means finding a feature of your subject that will interest your readers directly—whether or not it interests you—and beginning by talking about that feature. If the process you are describing can save the company money, you may be wiser to state as much in your technical memorandum before moving on to the special qualities of adhesion and flexibility that interest you. Your full report will have a better chance of being read with favorable eyes.

- Remember that enthusiasm is infectious. People who talk about their subjects with animation manage to captivate readers or listeners who never gave the topics a thought before. Those who don't, don't. I am reminded of a project director who was mumbling through the details of the work at hand to a group of company executives. His facts were all there, and the procedures were clearly spelled out. At one point, the director looked up from his notes and bleated at his audience, "It's really very exciting." Whereupon a restless vice-president roared in response, "Then for God's sake, get excited about it, man!"

- A generalization that is true and important to people will attract their attention and make them agreeably disposed to read on. Suppose you are writing your superior a memo on an industrywide conference you attended. If you begin with a hint about the priceless pearls of wisdom to be picked up from competitors, you will entice him or her to read more about this intriguing conference. But, if you start, as most do, by saying that this was a conference on A, which met in the city of B, state of C, for the purpose of discussing D, and then go on to say how it differed slightly from last year's meeting—can you blame your reader for slipping the memo quietly away, unread, into a "conference" file of other such memos?

- The language of persuasion is *not* necessarily "persuasive" language! In fact, expressions such as, "I *know* you'll be interested to hear that . . . " tend to put the reader on guard. How can this writer know what interests me? Why should he or she assume such intimate knowledge of my mental makeup? The poor writer is handicapped before he/she even gets to the subject of all this interest! No, the language of effective persuasion is, above all, language that is a *pleasure* to read. The reader should not have to expend effort to get through it. There should be no fuzzy verbiage, no irritating jargon, no self-conscious hedging. Sentences should be reasonably short, with only one thought per sentence. That single criterion will help you banish muddling complexities from your style. Grammatical mishaps that make the reader pause and wonder "why that doesn't sound right" must of course be eliminated. (I will discuss the most common of these in a future column.) Finally, the language should be appropriate to the readership. No unexplained scientific terms for nontechnical people should be on your list—and, above all, no suggestion of "talking down" to anyone. Assumed superiority is one of the greatest enemies of persuasion!

The last, most difficult, requirement of persuasive writing is simply to know when you have said your piece—and stop!

Communication ≠ Information

Part 1

Let's look at that word, *communication*. It stems from the Latin word, *communis,* which means "common." It has to do with what is common to all of us—our humanity, if you will.

To communicate means far more than *to inform*. Yet too often in business we confuse the two. We *inform* people of the facts as we know them, and consider that we have *communicated* with them. And then we wonder why we don't get more response.

The informer provides data. Period. The word has even acquired sinister overtones, as we think of police "informers" or foreign agent "informers." Perhaps the word became apt for such officers precisely because it suggests a lack of humanity. An encyclopedia can inform, a data sheet can inform, a computer can inform. But only a human being can communicate.

The great communicators are those who reach out into the common ground of humanity and seek to touch as many human needs as possible. Not just the need to know, or "be informed," but other equally powerful needs that propel us to work, to play, to make decisions, to choose one method over another—to promote one person over another. Communication touches all these needs.

What are they? I'll give you my list in a minute, but first, take a look at yourself. What are *your* needs?

To start with, why do you work? Yes, for money, but is that all? Or is there something in your job, your profession, that you really enjoy, for its own sake? Something that fills a need, makes your day exciting, worthwhile, something to think about and talk about when the workday is over? Something that's worth as much—maybe more—than the money you make?

Do a spot of self-research with me. Take out a sheet of paper and write down all the things you would do if you had all the money you needed. Then, next to each one, write down why.

Your "why's" will give you a list of *your* most important needs. Perhaps they include:

Companionship
Approval
The pleasure of a job well done
Mental stimulation
Discovery and broadening of your knowledge
Aesthetic satisfaction.

That's my list. You may share some of these needs and add others. Together, our lists probably comprise most of the basic human needs. These are the needs that the good communicator determines to answer.

Now—what happens when you reach out like this? After all, it's a fair amount of work. Why go to so much trouble?

What happens is the alchemy of communication. Instead of one person actively giving information and the other passively receiving it, communication sparks activity in both. The reader mentally reaches out to the writer in response.

The reader makes a distinct effort to understand and even sympathize with the writer. When the writer touches and seeks to fill the reader's needs, the reader responds in kind.

Think of your own response to letters that were honest, courteous, friendly, and clear. Now think of your response to letters that just threw some information at you and left it at that.

We write because we hope for a favorable response. So—real communication is worth the effort.

How do you go about answering all these deep human needs in a simple, practical business letter, memo, or report? You start by *being aware* of the needs and trying hard to fill them. Most people never get that far. They're too focused on *informing* to think about *communicating*.

Then, you consider each need of your list.

Companionship. You will write a friendly, personal letter or memo that suggests a smile and a handshake, not a computational buzz. You'll use the language and the tone you would use when speaking to a friend. You will let your personality show through and share it with your readers.

Approval. You will approach your readers with respect, making them feel that you honestly care about their response and value their opinion.

The pleasure of a job well done. You will write a memo that is complete, understandable, and reasonable. If you have a suggestion to make, show your readers that it will indeed help *them* in *their* work (not just you in yours).

Mental stimulation. Assume your readers want to be stimulated. Share the excitement of your findings or proposals with them. If you're writing about something that interests them—as you should be—assume they're willing to put some effort into thinking about and understanding it. And give them the information they need to do so.

Discovery and broadening of your knowledge. Like you, your readers want to know more, especially about things that affect them and their work. Tell them—but in clear, simple language that they don't have to decipher. Give them the background knowledge they need to understand what you're about to tell them. Write coherently and logically.

Aesthetic satisfaction. Make your memo or letter a pleasure to read. Use words with care. Don't keep hammering away at the same sentence structure. Don't overfill your page or your paragraphs. Read it all over to hear if it *sounds* good to you.

Next time, we'll look at a memo that fills all these needs—and one that doesn't. The second is a prototype of the memos that cross your desk every day, without making your life any better for it. The first is the kind that gets results!

For now, just concentrate on remembering the needs that a good piece of communication tries to fill. You'll find you're already writing differently. □

Communication ≠ Information

Part 2

Last time I suggested that communication is *reaching out,* rather than just informing, *filling the reader's needs,* rather than simply stating yours. Now let's look at two renditions of the same memo—one written according to traditional form, the other written to communicate. This memo is a report to your boss on a conference you attended on management techniques.

The old way

Fill in the heading with a minimum of information. After "Subject," just write, "XYZ Conference, April 5, 1983."

Begin by announcing what you're writing about, using as many words as possible, preferably in the passive voice. For example:

> On April 5, 1983, a trip was made to Appleton, Wis., by B. Shore and J. Abrams, for the purpose of attending the XYZ Conference on management techniques. The scope and subject areas covered by the XYZ Conference are described herein.

Go through the conference chronologically, giving a general idea of each topic discussed.

Close by mentioning that the traffic control techniques suggested in one lecture sounded worth trying—but don't commit yourself.

Append a resumé of each speech or session.

That's a typical "trip memo"—complete, factual, and boring.

The new way

Begin *talking*, in your mind. Set up a dialogue with your boss. Consider *his* interests, priorities, wants, needs, character, and present state of mind. And write down—or dictate—the dialogue exactly as you hear it. Here's an example:

You: I'd like to tell you about the XYZ Conference on management techniques that Bill Shore and I attended on April 5th.

Boss: Ah. . . well, I'm pretty busy right now. Was there anything that would really interest me?

You: Yes, indeed. Mike Barker described a new way of controlling traffic that I think could save us a lot of time and irritation—particularly during times of heavy deliveries. Barker is vice president of sales at the Brass Band Corp.

Boss: What does Barker suggest?

You: *Tell him—in a couple of sentences.*

Boss: Yes—that might work for us. Was there anything else we could use?

You: *If so—outline the interesting talks briefly.*
If not—say the conference was routine, but you're appending a short summary of each talk should he be interested.

Now, write your memo from the dialogue. Make your heading information *complete,* so you don't have to repeat it in the text. After "Subject," write what you first said to your boss:

> J. Abrams and B. Shore's report on the XYZ Conference on Management Techniques, held in Appleton, Wis., April 5, 1983.

Now you can use your first paragraph to *get the reader's interest.* Answer his needs or questions first, right up front. Never mind the chronological order of the conference. Go back to your dialogue and begin your memo with your answer to your boss's main question: "Was there anything that would really interest *me*?"

In this memo, you begin with Mike Barker's great ideas for traffic control. Tell your reader succinctly what Barker's technique is and how it might save the company time and irritation.

The remainder of the memo is the remainder of your dialogue. If there's more of interest to tell, tell it briefly. If not, say so and append your summary as a separate sheet.

The "need" test

To see how well these two memos rank as pieces of human communication, we can match them against the list of human needs (see "Communication ≠ Information, Part I," *Tappi Journal,* April 1983, p. 98).

Companionship. In the first memo, the writer made every attempt possible *not* to let his personality show through. In the second memo, the writer was trying to create a companionable meeting on paper. He was literally speaking to the reader, imagining the reader there with him. The reader would feel that personal act of reaching out to him.

Approval. The first memo shows no interest in the reader's opinion. It's just a catalogue of facts. But the new memo says, "Look—I value you greatly. I'm anxious to hear what you think about this new idea for traffic control; that's why I'm telling you about it right up front."

Pleasure of a job well done. Drab and lifeless, the first memo is not *itself* a job particularly well done. It doesn't give the reader any conviction that it will help him get *his* job well done. But memo number two does both. In its presentation of the conference from the point of view of the reader's interests, it proves that the writer did a first-class job as company representative there. And, in its enthusiastic, informative discussion of the new traffic control techniques, it shows the reader a way to get his own job done better.

Mental stimulation. Memo number one gives the reader nothing to think about, unless he reads diligently to the last line. And that one's vague. Memo number two offers a solution to an annoying problem and possibly adds some further interesting insights gained at the conference. It starts the reader thinking.

Discovery and broadening of knowledge. Memo number one gives the reader information. Memo number two turns that data into practical knowledge, as the writer explores the possible company uses of conference deliberations.

Aesthetic satisfaction. The old memo is dull, shapeless, and boring. The new one flows with natural speech and takes on the form of a satisfying conversation, as it answers the reader's unspoken questions as they arise.

Simple as it is, memo number two is a super communicator. By filling the reader's needs, it gets you what you want: complete attention, approval, and a good chance that your suggestion will be tried.□

Science and language—the search for elegant simplicity

It has been said that aspiring writers of fiction should do their apprenticeship in science or sports reporting. The reason? These two disciplines call for absolute adherence to the principles of the craft of writing:

Clarity
Communication
Logic
Originality

The science or sports writer cannot afford ambiguities. He is not trying to steep his readers in a mood. Rather, he has some precise information to convey. Furthermore, that information cannot be given any old way. The reader wants to know what happened, when, and what caused it. Finally, both science and sports have been so beleaguered with jargon that a fresh, incisive report is difficult to create. Most writers fall back on the cliches that have lost all precision of meaning through overuse.

You can make your papers or reports clear, exact, useful, *and* a pleasure to read if you follow a few easy guidelines.

1. Simplify. Complicated facts can be related simply. If you're shaking your head as you think back to the last, complex lab report, read Einstein's single-sentence postulate on his "special principle of relativity":

> If a system of coordinates K is chosen so that, in relation to it, physical laws hold good in their simplest form, the *same* laws also hold good in relation to any other system of coordinates K' moving in uniform translation relatively to K.

That's the theory of relativity! Not a single word of jargon; not one phrase or concept that the general reader cannot grasp. Contrast the master's clearly stated principle of relativity with a "technical" description of group therapy:

> As a result of the verbal interaction of participants, there evolves a collective understanding which in turn stimulates further co-individual responses, coextensive with the newly emerging group behavior . . . .

Think of translating the events or theories you describe out of their specific technical world into clear, uncomplicated English. Then your sentences won't need therapy!

2. Visualize your readership. A select group of specialists won't need explanations for much of your material. You can concentrate on giving them an accurate, full presentation. A more general, less knowledgeable readership will want a minimum of technical language and will need the facts to be made vivid by the use of familiar comparisons. In other words, the wider your readership, the less technical, more creative your work must be. It is not enough to communicate a fact; you must make it understandable and interesting within the reader's frame of reference. Look how Kenneth R. Miller introduced the complex process of photosynthesis, making it clear and interesting to all readers.

> The earth is a planet bathed in light. It is not therefore surprising that many of the living organisms that have evolved on earth have developed the capacity to trap light energy. Of all the ways in which life interacts with light the most fundamental is photosynthesis, the biological conversion of light energy into chemical energy.
>
> *Scientific American,* October 1979

Miller began with an image everyone could enjoy. He made photosynthesis understandable as a logical consequence of that image. Gradually, he moved from the familiar to the unknown, explaining scientific facts through a world the readers knew.

3. Describe the process step by step. One of New York's largest public relations firms tests would-be copywriters by asking them to write a press release for pliers! The candidates have to assume that pliers have just been invented. The press release must explain how the tool works and why the reader will benefit from owning one—and it has to be arresting enough to hold the reader's interest all the way through. Scientific writers would do well to give themselves this little exercise. Take an instrument and explain how it works, logically, building fact upon fact. Set aside an hour for this project, and you may save days of effort in your technical writing over the next few years. Remember, you are trying to impart information to a reader who knows nothing about the subject. And you will make that information interesting by relating it to the world the reader does know and care about.

4. Avoid scientific jargon as much as possible. Remember Einstein's introduction to his theory of relativity. It was clear, concise, readable English. Of course, there are times when you will have to use a technical word, simply because there is no general English expression for it. But don't fall into the trap of superspecificity! Because of their precision training, many scientific writers believe each word or concept they use must have a single, specific meaning. Since many words that formerly belonged to science have been taken into the general language (e.g., *magnitude* from astronomy, *force* from physics), these words have lost their original single focus and have acquired a more diffused meaning. Consequently, scientists keep raiding Greek and Latin for ever more specific words to append to newly discovered phenomena. They come up with a whole new language filled with mouthfuls like bioluminescence and telesthesia that nobody else can pronounce, much less comprehend! Keep these words to a minimum. Sometimes it is better to be less specific and more widely understood.

5. Never forget that your purpose in scientific writing is the communication of knowledge. Clarity, easy English, a logical presentation, and an understandable frame of reference are all tools to help you reach this single, all-important goal. If you keep that goal in mind, you will probably achieve it with very little effort. Read your piece over to see whether you would understand it if you had to face it cold. And, as you remind yourself that communication is everything, think of the German symphony conductor Karl Richter. During a rehearsal, Richter lost his temper with an arrant British trombonist and yelled, "Up with your damned nonsense twice or once I will put , but sometimes always by God never!" Richter's grammar could have been better, but, according to Lord Mancroft, who tells the story, "The trombonist got the message."

Technical writers: Loosen up

Why does writing sometimes seem so much more difficult than talking? It's all words, words, words... yet when we have to write them down, we often freeze. One reason, I suspect, is that we write in a vacuum. We don't imagine the reader; we don't mentally supply his responses or cater to his rate of understanding. So, faced with a subject and a demand for a competent written rendition of it, we cast around for rules and principles to get us through.

Waste no words, but...

This is fine, so long as we remember that all these rules are merely means to an end. The goal of any type of writing is simply *to reach the reader,* to make him take interest in your words and ease his understanding of them. A number of principles can help you reach this goal, among them a rule favored in this column: *Cut out wasted words.* This rule is designed to steer the writer away from verbosity and other forms of loquacious pomposity that muddy the message instead of translating it. But, again, like all the principles of writing, this one can get distorted when exaggerated and made an end in itself.

Here is an example. Like many technical writers, the producer of the following passage was no doubt anxious not to overwhelm his reader with the wealth of information he had to transmit. He determined to waste not one word and set it all across in the minimum space and time:

The accepted waste paper—grocery waste, newspapers and magazines—on arrival at the inlet to the machine is measured and electrically funneled onto a conveyer which feeds into a perforated drum, where the waste migrates down, the small solid rejects passing through the perforations, onto a horizontal conveyer for manual removal of large non-reusable materials (wooden crates, plywood, wire, plastic, etc.) that have not passed through the drum.

Whew! The strain of the writer cramming all that into one sentence can only be equalled by that of the reader trying to digest it. Note that the sentence is perfectly grammatical. No syntactical slips or dangling modifiers. *No extra words.* The writer could not be accused of verbosity, even though his sentence leaves your head spinning with unconnected words.

Write to be read

What happened? The writer didn't think like a reader. Had he done so, he would have soon realized that his offering was indigestible. Reading is sequential; we take in one concept at a time. If possible, we *imagine* the thing described. We are, literally, trying to get the picture. We can only put it together piece by piece.

In this passage, the writer is describing a machine—which we try to visualize as we read—and its operation. The reader should be led to see the action and neatly follow its progression.

Instead, he starts out with at least seven images that don't go anywhere: *accepted waste paper, grocery waste, newspapers, magazines,* their *arrival,* an *inlet,* the *machine itself,* and any other images these words trigger all have to be juggled in the reader's mind before he gets to a verb that tells him what happens to it all. The reader may pause to wonder how it got there (I did!), but before he can stop to think about that, the verb pulls him up short. *Is measured.* Where is it measured? How? For what? Measured for weight or bulk? These are questions we might really like answered—but we have to put them on hold, too, for our writer is off and running. We have to drop the concept of *measuring* and pick up *electrically funneling.*

Again, the reader may pause to wonder, as he tries to visualize the operation. How does the funneling apparatus work? Where is the power? But, even as he tries to make himself a picture based on his own knowledge, his eye leads him on to a *conveyor* and a *perforated drum* that *rotates.* He has to abandon the unformed picture and the questions it inspired and start hurling conveyors and perforated drums into the grab bag of images that is overflowing in his mind.

And that's just in the first clause. From then on, the reader's imagination has to leap from one conveyor to another—this one horizontal (as distinct from the other one?)—from small rejects passing through perforations to large ones that surrender themselves to manual removal and that comprise crates, plywood, wire, plastic, etc.

In short, it's much too much. The reader can't possibly take in all this information in a single sentence. "Taking in" means thinking about, wondering, asking oneself all those questions that the reader has had to put on hold. Of course, the questions don't always move out of focus. The reader finds himself still musing over one concept while his eye has moved on to another. And so his mind misses what his eye sees.

Furthermore, a sentence has an inner hierarchy. The main clause is the most significant, the others, more or less important depending on their position and grammatical structure. But most of the information in this sentence is of equal importance. The measuring of the waste is no more significant than the filtering out of small rejects — but the sentence structure makes it seem so. The structure of writing should reflect, not distort, the relative importance of the concepts discussed.

A pruning process

What to do, if you find yourself writing like this? First, remember the goal of *readability*. Imagine a reader facing your paper for the first time. Don't be afraid to use as many words and sentences as you need to give him the chance to absorb and understand your meaning. Let the flow of your writing reflect the flow of the event described, with pauses between the actual phases.

By all means, prune the wasted words—the cliches, paddings, and fuzzy expressions that transmit nothing but vagueness. But then, loosen up. A good rule is to *give each significant concept a sentence of its own*. For example:

The machine accepts grocery waste, newspapers, and magazines. First, the waste is measured at the inlet to the machine. Then, it is funneled onto a conveyor which will carry it into a perforated rotating drum. As the waste moves through the drum, small solid rejects pass through the holes in the sides. The remaining material migrates onto a horizontal conveyer. There, technicians separate the large rejects, such as wooden crates, plywood, wire, plastic, from the reusable paper waste.

The writing takes you step by step through the operation. Using just *ten* more words, we've made sense of wastes—without a wasted word!

Are you an impressionist or an expressionist?

"I have little patience with scientists who take a board of wood, look for its thinnest parts, and drill a great number of holes where drilling is easy."—Albert Einstein

If you want to impress—with ease—talk or write nonsense. So concludes J. Scott Armstrong in a report on studies of audience and reader reaction made by himself and others. The results indicated to Dr. Armstrong that: "An unintelligible communication from a legitimate source in the recipient's area of expertise will increase the recipient's rating of the author's competence."[1] Or—people applaud gobbledygook, so long as it's grabbed by an "expert." Some of the evidence cited by Dr. Armstrong follows.

- An actor, who looked and sounded distinguished, was sent to speak to a group of experts on a subject about which he knew nothing. His made-up biography was impressive. His talk was sheer nonsense, contrived into a semblance of coherence through false logic, contradictions, and irrelevant references. The audience of psychiatrists, psychologists, and educators said it was a clear, stimulating lecture.
- A survey of reader reaction to management journals showed that those journals with the lowest "readability" were considered the most prestigious.
- Clearly written papers were rejected by widely read trade journals. When the authors rewrote them to make them incomprehensible, they were accepted.

These are indeed sobering discoveries, for they show that language and meaning are growing farther and farther apart. We are back to tribal rites round the campfire, where the chant itself is senseless. All that matters is that you sit at the "right" fire, with the right tribe, mumbling the "in" mumbo-jumbo.

The studies take us to a fundamental question facing all users of language. Are we speaking (or writing) to express or request knowledge? Or is our purpose simply to *impress* people, without too inordinate an effort? Do we want to "drill a great number of holes where drilling is easy"?

Before you read on, think back to some recent letters and reports you wrote or a technical paper delivered to your peers. Were you trying to *express* your knowledge clearly—or *impress* people with its vastness and acuity? If you are an "impressionist," you may be doing yourself a disservice by striving to write well.

Seriously. Clarity and honesty rarely serve impressionists' ends, the studies show. Rather, the impressionist would do much better to follow these rules:

- Use as much technical jargon as possible; peruse every normal English word you have written and see if it can't be turned into a "techno-verb."
- Eradicate all short and simple sentences from your paper. Seek the compound-complex brand only, and add as many qualifying phrases as you can muster.
- Never leave a noun or verb unmodified. Plug in adjectives and adverbs—preferably two or three at a time, and polysyllabic, please—wherever they will fit.
- Make frequent and liberal use of the following words and expressions, and don't be afraid to repeat them. Each can serve excellently to mean—nothing.

area	it happens that
aspect	larger in size
case	lower in height
character	along these lines
circumstances	the nature of
factor	of the . . .variety
field	the way to . . .is
function	as far as . . .is concerned
respect	
type	
situation	

- If you've been following this column, go back over past articles—and turn every rule around.

This is not a joke. Working hard to write clearly when you really want to write impressively is like struggling painfully to swim—from your place in the sun on the sand. You don't get anywhere, and it isn't any fun.

But if you deplore the easy way when it's the false way . . . if you'd rather not spend your time drilling holes in thin pieces of wood, however much praise comes from it—then turn those last five rules around. And work hard for honest, expressive communication, not impressive doubletalk, in your place of work.

A good starting point is grammar. It doesn't have to be "perfect"—none of us is writing for school grades—but it does have to be *logical*. With so much fuzzy "impressionist" language around us, we have to check to see if the fuzz has covered our own thinking. For most of us, grammatical or syntactical pitfalls are the result of an *illogical* use of words. We may not have trouble deciding whether or not to use "ain't" or "he don't." But we can be stymied by such monsters as:

- collective nouns (such as *number* or *couple*) that appear in the singular but indicate a group or quantity. Let logic decide which verb you use. You wouldn't say, "The couple *was living* in separate houses," since here, you're thinking of two people. But you could say. "That's a nice couple," for there, you're thinking of the couple as a single unit.
- either/or and neither/nor. In school, we learned that they always take a singular verb. But what about sentences like: "Neither the President nor his Cabinet members *was able to* attend"? Think *logic* again. You're talking about several people. And the plural noun, *members*, is closer to the verb. So, of course, you write instead: *were able to attend*.

We'll discuss logic and language at greater length in the future. Meanwhile, let's try to make this the age of *ex*pressionism!

[1]Armstrong, J. Scott. "Creative Obfuscation," *Chemtech*, May 1981, 262-264.

How to overcome writer's block

It's one of the most frustrating moments of the day, you sit down in front of a pad of paper, typewriter, or word processor. You stare at it, willing that blank space to come alive, to make contact with your brain and turn your half-formed ideas into organized, communicable thoughts. It doesn't happen—you wait. You go to the water cooler in hope of irrigating the sluggish brain cells. Nothing. You get a cup of coffee, your heart speeds up a bit, but there's still nothing on the page.

You are suffering from writer's block, an occupational hazard of everyone who ever has to write. Since I know it only too well myself, I decided to analyze it and try to work out a solution fo it.

First, I found that writer's block is usually a problem of *organization*. When I was staring at the paper or screen, all sorts of thoughts on the topic would come tumbling into my brain, but they were so quick, incomplete, and disconnected I could not possibly write them down.

Eventually, my brain would tire of these senseless acrobatics and move on to something altogether different, such as "Wouldn't it be nicer to be skiing right now?" And off it would slide into pleasant meditations on skiing in heavy powder or making better parallel turns. But the page before me would remain blank.

Once I saw that writer's block was basically disorganized thinking, I worked out a simple technique to overcome it. And the great news is, it works. Writer's block is now part of this writer's history. Here's what you do.

Organize your thoughts

At the top of the page, write your topic. Then write "PURPOSE" and note *what* you want to get across. Next write "METHOD" and note step-by-step *how* you plan to do it. Finally, write freely on each step. That's all there is to it.

But your problems may not be entirely over. Before you start writing freely, you may have to deal with: exhaustion, insufficient knowledge, or sustained muddleheadedness.

Exhaustion. Writer's block saps your energy–at least it does mine. When I first tried this technique to break writer's block, I was exhausted from staring at the empty paper. I could only make a few notes on my purpose and method before I left it. The next day though, I followed my notes and wrote the piece quite easily.

Remember, writing is extremely demanding. It involves steady, concentrated thinking to a purpose. If you're too tired to write, put your work away, close your eyes, and take a break. You'll write faster and better later.

Insufficient knowledge. You may need to do some more research. You'll see what and where as you map out your method. If you find yourself hesitating or worrying over a subtopic, just put a star by it and go on. Determine to check that topic before you start writing. This quick technique can save you hours of wasted doodling, as it will show you quickly and precisely what you need to check.

1. The best way to tackle writer's block is to beat it before it beats you. At the top of your page, map out your intentions, including purpose and method to accomplish your goal.

WRITER'S BLOCK

PURPOSE: To show people how to overcome it.

METHOD:
1. Describe writer's block
2. Discuss reasons it occurs
3. Give technique for overcoming it
4. Illustrate with an example

Sustained muddleheadedness. Don't laugh. Sometimes the state of writer's block is difficult to escape. The writer seems almost determined to stay stuck.

Let's suppose I'm blocked on this column. I dutifully write out my notes on purpose and method as in **Figure** 1. Now I should start writing on Step 1 under METHOD, but I won't if I'm in a state of sustained muddleheadedness.

Instead, I'll start wondering "Is this *really* the best way to proceed? Would it be better to devote this column to a description of writer's block and tackle the technique for overcoming it next time? Or . . .?" The possibilities are endless; the result is always the same, an empty page.

The cure for sustained muddleheadedness is perfectly simple. Refuse to consider your own questions! Just write strictly on each point you noted under METHOD. Of course you'll have some editing to do; you may throw out whole sections and add others. But at least you'll have something to edit! Happy writing.

A universal outline

How many times has someone talked to you clearly and interestingly about a project, only to follow the discussion with a boring, unfocused, unclear memo? Worse—how many times have you written such a memo yourself?

The cause of this sudden inability to communicate on paper is often an obsession with *form*. We pull out a sheet of Memo Paper and start worrying. How do you set up a "perfect memo?" What's the right format? What goes first, last, and in between?

If you're worrying about form, you can't concentrate on content; the brain can focus on only one thing at a time. So, the memo you finally write can turn into a catalogue of details about the project, with the core and the matter somehow missing.

You can break out of this tyranny of form by using a single outline for everything you write. This outline will produce letters, memos, and reports that are clear, complete, and interesting to your reader. Using it, you'll begin with the information that matters most to your reader and then follow with details for him to examine as he wishes.

If your company has a certain format for technical memos or reports, you can use this outline to write each section and then to work them all into a coherent piece.

Part 1: your purpose

To produce a focused, well-organized, meaningful piece of writing, you must know exactly why you are writing it. The first step of your outline is to note your *purpose* for writing the memo, letter, or report.

Don't confuse purpose with *subject*. The purpose is not the topic you're addressing. Rather, it's your aim, your *reason* for writing this particular message to these particular people.

To note your purpose, write "I want to. . ." and then put down how you want to affect your reader(s). "I want to persuade Bill to come to the meeting." "I want to make the Executive Committee understand the significance of our work on the drying project."

Part 2: the imaginary dialogue

Now conduct an imaginary dialogue with your reader, or a typical reader if there are many. Imagine the reader asking you four questions. Your answer will form the outline of your piece. Reader's Questions:

- *What* do you want to tell me?
- *Why* would that interest me? (Tell me how this affects me.)
- *How* would it do that? (Tell me more about it.)
- *Anything else* I should know.

An example

Say you are writing a request for funds for a new machine. The request goes to the Capital Appropriations Committee, which is composed mainly of people in top management.

So many capital requests produce yawns and the shaking of heads because they begin with a lengthy, highly technical description of the history and problems of the work at the lab and the machines now in use. Much of this information is either incomprehensible or peripheral to the executive reader, who simply wants to know *what* you're asking for and *why* it would benefit the company to give it to you.

You can write a capital request that will be read and carefully considered by following the universal outline:

PURPOSE: I want to persuade the Capital Appropriations Committee to approve our request for funds for the XYZ machine.

IMAGINARY DIALOGUE (with member of committee):

COMMITTEE MEMBER: *What* do you want to tell me?

MEMO WRITER: C lab needs an XYZ machine.

COMMITTEE MEMBER: *Why* would that interest me?

MEMO WRITER: The machine will make the company more competitive in . . . and should save us X amount of dollars over five years.

COMMITTEE MEMBER: *How* would it do that?

MEMO WRITER: It will give a higher-quality product (explain briefly) and will increase our productivity in three ways. (List them.)

COMMITTEE MEMBER: *Anything else* I should know?

MEMO WRITER: XYZ machine is more efficient and easier to operate than the present one. (Explain.) It would require on-site training of approximately one month. Further details on the machine and our plans for its use follow.

That outline will produce a good, clear capital request. The outline has worked for hundreds of people on many writing tasks. It clarifies thinking as well as writing. Try it next time!

Beginning, middle, and end

Part 1: "How to begin"

A whole is that which has a beginning, a middle, and an end.
- Aristotle, *Poetics*

Why is it always so difficult to begin a piece of writing? Seasoned writers who tear through page after page once they've got started—still have trouble getting started! Why?

Probably because that first sentence has an outrageous amount to do in a few words. To induce the reader to read beyond the first sentence, the writer must use it to say:

This is who I am, and this is who I think you are. This is what I have to tell you. And this is why you will want to hear about it.

No wonder first sentences cause momentary paralysis! But you can start well without stalling if you just remember *The Three T's*-TONE, TOPIC, and TANTALIZATION.

Tone

The tone of your writing says clearly: "This is who I am and this is who I think you are." Tone is simply the reflection of your attitude toward yourself and your reader.

Look at the opening sentence of this column. The phrasing of the question reflects, I hope, my feeling that you and I are together in this: We have both experienced the difficulty of knowing how to begin. But if I had written, "Why do you always find it so difficult to begin a piece of writing?" the friendly tone would disappear, to be replaced by a more distant, socratic one, agreeable only to those who like to be treated as students. Most adults don't.

To get the right tone for your opening, you must have the right attitude to your reader. That attitude is: **friendly respect**. Never vary from it, and your tone will always be right.

Topic

A good opener gives the reader an idea of your topic. It says briefly, "This is what I have to tell you."

When Abraham Lincoln wrote, "Fourscore and seven years ago our fathers brought forth upon this continent a new nation. . .," he told the listeners at Gettysburg that he was going to talk about loyalty and patriotism ("fathers") and the **whole** nation that he wanted to restore.

And when E. M. Forster began his great novel, *Howard's End*, he left his reader in no doubt as to what was to follow. He wrote simply, "One may as well begin with Helen's letters to her sister." And then he did!

Let your first sentence guide your reader to the topic you're going to address. Don't try to tell it all. That's the job of the rest of the piece. Just give the reader a few words to show him where you're going.

Tantalization

"This is why you will want to hear about it." Your opening sentence should inspire your reader to want to read on. It can do so by hinting that you're going to speak to a need or interest of his.

If you're writing on a topic of general interest or concern, a pointed question can be a good tantalizer. I began this column with a question to encourage you to wonder about the answer and to read on in the hopes of finding it.

But statements can be equally tantalizing when they are short, direct, and unexpected. Consider Melvill's famous opener to Moby Dick: *Call me Ishmael.* Who could read those three words and **not** go on to find out who *Ishmael* is and why on earth we should call him that?

Before you start writing, picture your reader—including his work, position, concomitant interests, and anything else you know about him. Once you can see him, try to catch his interest with something that relates to your topic. Ask him a question that you think he'd like to have answered. Mention a part of your topic that you know concerns him.

If you can't find anything specific to him, think of something about your subject that interests you as a human being—not just in your professional capacity—and try that out on him. When you find an interest or need, hint that he'll find out more about it in the paragraphs that follow.

You see, it's really not so hard to begin! Adopt an attitude of friendly respect to your reader, and you'll have the right TONE. Then give him an idea of the TOPIC you're going to address—just enough of an idea to TANTALIZE him! And you'll have an opening sentence that does its impossible tasks perfectly.

Beginning, middle, and end

Part 2: "The middle"

Aristotle's separation of a whole into a *beginning, middle,* and *end* points to the different functions of these three parts. Think of the middle part of a piece of writing as a journey on which you guide your reader through your topic. Imagine that you're actually taking him on a tour, stopping at the highlights. You will keep him moving and stop him from straying off into his own thoughts if you think *flow, focus,* and *flashpoints.*

Flow

Each sentence is a complete entity, a small island that needs a bridge between it and the next one if the reader is to move along with your thoughts. Look at this paragraph:

> "A tentative schedule for the meeting is attached. Participants should submit their proposals to this office by June 28. Schedules must be finalized by July 5."

The three sentences are three separate statements. Although the connections among them are not hard to imagine, the reader shouldn't have to supply them. A traveler doesn't have to build his own bridges to cross every little river. As T. A. Rickard once stated: "This is the first great principle of writing—economy of effort on the part of the reader."

To minimize the reader's effort and allow his reading to flow, set up the bridges for him. Check each sentence against the one preceding it. Have you shown the connections between your thoughts—or did you just make a series of separate statements?

If the writer of the above paragraph had connected his thoughts, he would have written something like:

> "I have attached a tentative schedule for the meeting. Since we must finalize it by July 5, we are asking all participants to submit their proposals to this office no later than June 28."

Think flow instead of statements, and use connectives. Here are some words that can sometimes provide the necessary bridge:

- And
- But
- Therefore
- Furthermore
- However
- Consequently
- Also
- On the other hand
- Finally

Focus

A paragraph is held together by a single focus—a topic or idea, to which you point in the first or second sentence of the paragraph. That sentence is the guidepost to the paragraph. It shows the reader what you're going to talk about.

Once you've announced your paragraph's topic, go on to talk about it! If the first sentence was "The high viscosity of this liquid is a problem," don't go on to discuss the temperature or quantity of the liquid! Talk viscosity.

This may sound obvious, but check back over some of your past drafts. This is what often happens. We write, "The high viscosity of this liquid is a problem," and the last word, "problem," stays in our minds. It may lead us to reflect that quantity is also a problem, perhaps one that could be solved more quickly and easily than viscosity. We suddenly get a great idea about reducing the quantity—and we're off and running. The paragraph veers into a full-scale discussion of quantity reduction.

Now, this creative, associative thinking is absolutely fine for the first freewritten draft. But when you come to shape the body of your work, be sure your reader gets what you promised in the opening sentence of the paragraph. Don't shift your focus.

Flashpoints

You want your reader to stop and look at the important points on your tour—the "flashpoints." Here are some ways to emphasize those points. Of course, emphatic devices work only if you limit them to the two or three points you want to stress.

Underlining. "We ask participants to send us their proposals no later than June 28."

Repetition of words. "This is a dangerous area. Do not enter unattended. Do not touch anything with bare hands. Above all, do not step off the walkways provided."

Repetition of sentence structure. "Getting to the site will be a problem. Finding the source of the power failure will be a puzzle. Correcting it will be the biggest challenge of your career."

Notice that the repetition also acts as a connective to the sentence whose words or structure it mirrors.

Remember, in the middle section of your work you want your reader to follow you with interest and with particular attention to your main points. Think *flow, focus,* and *flashpoints,* and you'll take the reader where you want him to go.

Beginning, middle, and end

Part 3: "The end"

This is the way the world ends
This is the way the world ends
This is the way the world ends
Not with a bang but a whimper.
—T. S. Eliot, "The Hollow Men"

And that is the way many letters, memos, and reports end, too. A dead, catch-all phrase—"Hoping to hear from you soon, I remain . . . Very truly yours," a whimpering warning—"I would be much obliged if you would attend to this matter at your earliest convenience," or a lame statement of the obvious—"This concludes the third quarter report of the X department of ABC Company."

Don't end your work with a whimper! Your final words are *very* important. As the last statement or question he has read, that final passage will remain in the reader's memory. It is your last chance to make him feel like doing what you want done.

We remember best what we read last. If it was humdrum or irritating, we tend to remember the whole piece as being that way.

Instead, determine to write an effective ending. You will do so if you meet three conditions:

1. Show goodwill in your tone and words.
2. Leave your reader with one significant thought.
3. Remind him of what he should know or do after reading your letter, memo, or report.

Signpost endings

If you have written a letter or memo that requires follow-up by your reader, the ending is the perfect place to remind him of what you would like to do. You can make this a simple, straightforward statement, such as this:

> Thank you for considering this proposal. I look forward to discussing its possible applications in your lab work when we meet on January 1.

But be sure to show goodwill. Don't hit the reader on the head with your reminder . . .

> I expect you to think this proposal over seriously and give me a list of possible applications for it in your lab when we meet on January 1.

Nobody likes to be told what's expected of him. You're much more likely to get good response if you close on a tone of friendly respect (the *only* tone you should ever use in any of your writing).

Emphatic endings

To be effective, your last paragraph should leave your reader with one important thought. Just one. If you try to cram in more than one idea, he may not remember any of them clearly.

You may wish to emphasize a fact or opinion that you hope will convince your reader to do as you wish. For instance:

> Devoting extra support to this project now may save us more than $30,000 over the next three years.

If your piece of writing contained a lot of information, it's certainly a good idea to summarize your main points. But a summary alone rarely makes a good ending, since by definition it contains more than one point. Instead of the last, make it the *next to last* paragraph. Then use your last paragraph to state the essential point or discovery. For instance:

> In summary, we studied five different particle sizes at three levels of capacity. At 85% load, we found a small tendency for carryover. At 100% load, the tendency for carryover increased. When we experimented with the particles at 110% of nominal capacity, the carryover tendency was twice what it was at 85% load.
>
> For every size of particle we studied, we found that carryover increased markedly with load size.

Provocative endings

When you're editing your work, look for any small points of interest that you could save for the last paragraph. I don't recommend leaving truly essential information until the end, as that would make you feel that you were stringing him along just to make a final splash. But you'll find you can often pull out a small example or an interesting figure to make an effective ending. Keep this in mind as you go over your draft.

For instance:

> So far, 593 people from 12 companies in 4 major industries have taken this seminar. Come and find out what's worth more than a day's work to all of them!

Remember, your readers take home the tone and message of your final paragraph. End with a bang, not a whimper!

Getting your ideas across

The difficulty is . . . not to affect your reader, but to affect him precisely as you wish.—Robert Louis Stevenson

In the last two columns, we looked at some of the dynamics of communication. Now I'd like to show you 10 ways to make these concepts work in your writing. Each is a simple writing device; together, they make a powerful recipe for persuasion.

1. Interest your reader

It may sound obvious—but how many technical memos have you read that rely on a dreary rundown of facts to convince you of the worth of doing something? As advertising mogul David Ogilvy says, "You can't bore people into buying your product."* It's the same with ideas. Think of a part of your idea that would interest your reader—even if it's not your main point—and open with that.

2. Tell the reader what's in it for him

Again, this may not be your main interest—but you can reach your reader with it. Suppose you want to institute informal weekly meetings, for just you and your immediate staff. When proposing this idea to your boss, you might write that these meetings will give him an hour of quiet, uninterrupted work every Wednesday at 9:00 a.m.!

3. Make your main point and stick to it

There are many things you can say about your idea—but you won't get it across to your reader if you say them all. You'll just muddle him. After you've gotten your reader's interest and shown him what your idea can do for him, go straight to your main point. (If you're lucky, your main point is also the point of interest and benefit to your reader! But circumstances are not always so obliging.) Once you've stated your main point, say all you have to say about it—before you discuss anything else. Indeed, there may be no time, space, or reason to say anything else!

Suppose you have found some new software that you think could save the accounting department thousands of dollars. Write down what the software is and how it could save so much money. Don't start comparing this particular software to a similar product from another company. Don't go into technical details about the program. You can do all that another time, after you have gotten your idea across.

4. Write one-to-one

Even if you are sending a memo to 50 people, each of them will read it alone. Pick one of the readers whom you want most to reach, and write as if you were talking only to him or her.

*David Ogilvy, *Ogilvy on Advertising*, Crown, New York, 1983.

5. Give testimony by people the readers respect

Testimonies by people who have tried your idea or something like it can be helpful. But it is essential that your readers think highly of the person you're quoting. Suppose you write, "Bob Jones says that his weekly meetings have boosted his staff's morale and productivity." You will help your idea if your reader considers Bob Jones a good, thoughful member of the company. But, if he thinks Bob would do anything rather than work, Bob's testimony will just nudge your idea toward the wastebasket!

6. Use specifics, not generalities

People pay more attention to them. If you write, "Bob Jones has a regular attendance of 60–70% at his weekly staff meetings," your reader will take note and remember it. He probably won't if you write, "Attendance at Bob Jones's meetings is high."

7. Don't use superlatives

People don't credit them. They won't take you seriously if you write "These meetings were undoubtedly the best thing Bob Jones ever did for his people." But they'll believe you and consider your suggestion if you say, "Bob Jones has noticed that the members of his staff work together with far less friction and misunderstandings since he instituted the weekly meetings."

8. Avoid analogies

People misread them. If you write, "Holding these meetings would be like giving the staff an hour of healthy calisthenics once a week"—your boss may well think your idea is to set up an exercise program.

9. Write short sentences and paragraphs

People in the technical areas of business often get itchy when professional writers tell them to use simple English words to describe complex technical subjects! The problem is valid and deserves consideration on its own (it will be the subject of a future column), but it's not the point here. No matter what you're writing about, you can write three short sentences instead of one long, convoluted one. And you can break up your paragraphs, so that your piece will look inviting to read. And then your reader will read it!

10. Number your points

If you have several points to present, you'll make them easier to read and more pleasant to look at if you connect them by numbers and then treat each separately—as I have done here.

Chapter 2
The Words

How to Choose Them, Use Them, and Put Them Together

These columns tackle the essential units of writing: the words themselves.

We begin with the *unnecessary* words that you can cheerfully eliminate. **Weeding out wasted words** and **Please—no bad language** show you how to free your writing of words that weigh it down.

What's wrong with a cliche, anyway? tells you why you should leave your cliches at home, along with your bedroom slippers.

Then come the "roadblocks"—the *difficult* words we all tend to mix up or misuse. A two-part series, **Roadblocks,** lists the most common stumpers and gives you ways to remember which ones you really mean. **Snarks and bojums** and **A checklist of problem words** supply the correct spelling and meaning for other commonly confused words. And **In small scale** highlights those words whose meanings can be changed by prepositions and shows you which preposition to use when.

The next two columns discuss effective use of the two major *types* of words—nouns in **Nouns and nounery** and verbs in **Being verbal.**

Finally, a four-part series of columns, **How to put power into your language,** teaches you how to choose the right words and join them to form strong sentences and paragraphs.

Weeding out wasted words

The project administrator fixed his staff with a steely stare, then looked back to his notes.

"It is of the most vital importance," he told them, "that if during preliminary discussions, the subcontractor has agreed to accept a change to any integral part of the original subcontract, with no additional claim to be made for extra cost, schedule stretchout, or other adjustment, the notice recording the aforementioned amendment be written in accordance with language which is to the point and without question."

"*What* was that?" one boggled engineer asked another.

"I think he wants us to be sure our subcontractors write their change notices in clear language."

"I see. Not like his, in other words," the engineer concluded.

The administrator's remarks—which should, indeed, have been "in other words"—are a direct quote from a company executive who was upset by the vagueness and confusion of the wording of subcontracts. We have reached the unenviable stage where, in our very complaints about business language, we can't make ourselves understood! Why? Because a tangle of muddled words ties up our thoughts—like weeds stifling every shoot that tries to bloom.

A "weed" in language is any word or expression that contributes nothing worthwhile to the sentence. Eradication of weeds, however, is not always simple. You may be filling your speech with weeds that do nothing but establish you as a member—or admirer—of the group that uses those words. A teenager gasps, "And what I like—oh no!"; a businessman declares, "My immediate reaction was, I can assure you, one of profound concern and dismay." Both mean: "I was shocked." Their words are merely the insignia of the group with which they wish to identify. If you earnestly want to sound like "one of the crowd," you must resign yourself to a tedious, unimaginative speech and vocabulary. But—if you want to be crisp, clear, and a pleasure to listen to, you'll have to expunge most business or technical jargon, for it rarely adds to the sense or the power of what you have to say.

Now, the most common weeds in business and technical writing are:

• Hedging expressions—words that the writer sprinkles into his sentences purely to weaken an already tentative statement. Typical examples are:

1. Excessive accumulations of negatives: "It is not unlikely that the bond will fail to hold above these temperatures" (meaning: the bond *may* fail to hold . . .)

2. "General" expressions, such as *as a general rule, usually, about*: "We can expect an approximate increase of about 3%" (approximate means "about")

3. Introductory phrases: "If I might add . . . " (you're going to, anyway, so why not just say it?); "Speaking personally . . ." (how else does a human being talk?)

There are almost as many ways to hedge as hedgers. To clear your speech of these stumbling blocks, ask yourself first if you can't just make the statement, with no restrictions. If you do want to hold back, one proviso is enough. If you've got an "about," you don't need "in the area of"; if you've said "may or "might," strike out "possibly."

• Redundancies—repetitive expressions that the writer plunks into his sentences in the mistaken belief that they will intensify the power or significance of his message. Curiously enough, both the hedging writer and the one who wants to drive his point home heap word upon unncessary word—for opposite aims. All they achieve is a cumbersome clutter of verbiage that obscures meaning altogether.

Redundancies that add nothing but tedium to your tale include:

if an only if, then and only then, etc.	= if, then
each and every man and woman	= everyone
each of these	= each
few in number	= few
consensus of opinion	= consensus
and so as a result	= and so
adequate enough	= adequate
and moreover	= moreover
but nevertheless	= nevertheless
never at any time	= never
results so far achieved	= results

Some less obvious redundancies appear in these common expressions in which the adjective duplicates a concept in the noun:

advance planning (there's no other kind of planning)

close proximity (you can't be in far proximity)

first priority (there's no second or third priority—despite political lingo!), and, by the same reasoning:

practical experience . . . *necessary* criteria . . . *integral* part . . .

major breakthrough . . . *new* recruits . . . *separate* entities

• Wordiness. Even if you're not trying to restrict or puff up a statement, you can get tangled up in excess words. Some cumbersome constructions found in business circles, followed by their happier, shorter alternatives, are:

in view of the foregoing (*therefore*) . . . owing to the fact that (*because*) . . . at the present time (*now*) . . . at an early date (*soon*) . . . of the order of magnitude (*about*) . . . for the purpose of (*to*) . . . in accordance with (*by*) . . . along the lines of (*like*) . . . for the reason that (*since*) . . . prior to (*before*)

So—take some time to do some weeding. You'll soon see your thoughts reflected by, instead of hidden in, your words.

Please—No bad language

A boy swears at his broken bicycle—and his mother scolds him for using "bad" language. A student admits, "It was *me*"—and incurs further outpourings of wrath from his sticklish teacher, who considers the faulty pronoun "bad" language. Neither the mother nor the grammarian was reacting to the sense of the offending words. No—they were put off by the words themselves.

I would like to consider "bad" language as just that: language that *puts you off.* In business communications, there are two main types of bad language. One annoys the reader by addressing him in an inappropriate tone. The other irritates him by obscuring the meaning you are trying to transmit. Both types can destroy your efforts at communication by putting your reader off before he even gets your message.

The wrong tone

With all the best intentions, you can alienate your readers simply by using an ill-advised expression that communicates something you never intended. For instance, you may begin a memo:

It has been brought to my attention that you . . .

This inflated opening suggests that the writer occupies an exalted state where he is brought matters of great pith and moment on a sterling silver salver. The reader, on the other hand, appears as a problem or event that has been submitted to the important writer's gaze. This is not an auspicious beginning. Instead, try to find an expression that immediately establishes a bond of humanity between you and your reader. One possibility is:

I was interested (glad, sorry, concerned) to know that you . . .

Here, the "I" and "you" are on the same level, and the "I" is expressing interest in the "you."

Other poor openers, frequently used, include:

Let me remind you . . . inform you . . . call to your attention . . . .

Again, these expressions suggest superiority and pomposity—neither of which is an endearing human quality. The writer is wagging a finger at the reader and telling him to sit up and listen, fast. People don't like that. You'll get a much more positive response if you use expressions such as, "As you may know . . .," or "I have learned that . . . ." Or, just state what you have to say without a preamble.

Check your drafts to see if any of the wording could imply a feeling of superiority, over-familiarity, or even subservience. Such a tone will interfere with your message. If you try to write your piece with as little fuss as possible, as if you were speaking, you'll find the "bad tone" disappears.

The detritus of language

The words I'm labeling "detritus" are those little ones that get between your message and your reader. They represent the disintegration of language, for they mean nothing. They're like a sandbank; the more you pile on, the higher and thicker the pile gets, until the message is finally obscured altogether.

Paradoxically, we tend to ram in these little, meaningless words when we're afraid the reader won't get our message without them. Insecurity breeds verbosity.

We may start out writing, "First, we must examine all the data." Then—we fear we haven't made the point strongly enough. "First *and foremost*, we must examine all the data." Perhaps this isn't clear enough? "First *and foremost*, we must examine all the *pertinent* data." But will this sound like an impossible task? "First *and fore-*

most, we must examine all the *pertinent* data *available.*"

And so it can mushroom, with not an iota of information added. *And foremost* can't make *first* any "firster." *Pertinent* dwells on the obvious; the reader knows you don't mean all the data in the world. *Available* is redundant; if it weren't available, you couldn't examine it.

Besides obscuring the message, words forming detritus are bad in another sense. Through their redundancy and repetition, they rob other words of their meaning. *First* loses its impact when you add *and foremost.* It becomes just a weak member of a well-worn cliché. Similarly, *pertinent* and *available* weaken all the other words in the sentence, for their presence suggests (erroneously) that they were needed to supply some missing meaning. The original lean and simple statement has been puffed up into a bland blur of syllables.

Look at your writing. Are there words that add nothing but syllables to your sentences? Have you started building a sandbank?

We use good language when we're good to language and our readers—when we approach them simply, honestly, and with respect.☐

Recognize and remove detritus

advance warning	=	warning
attached *hereto*	=	attached
as *in the manner described*	=	as described
basic fundamentals	=	fundamentals
circle *around*	=	circle
collect *together*	=	collect
connect *up*	=	connect
early beginnings	=	beginnings
empty *out*	=	empty
enclosed *herein*	=	enclosed
equally as good as	=	as good as
exactly identical	=	identical
it is frequently *the case that*	=	frequently
join (merge) *together*	=	join, merge
make mention *of*	=	mention
plan *ahead*	=	plan
refer (report, retreat) *back*	=	refer, report, retreat
resume *again*	=	resume
until *such time as* you can	=	until you can
very unique, *most* unique	=	unique

What's wrong with a cliche, anyway?

What's wrong with cliches? Nothing—when you're tired, emotionally distraught, or just taking a break from thinking. When you're having a long Sunday morning, stretched out in an easy chair and drinking a lazy cup of coffee—there's no reason why you shouldn't remark that you've been *dragged through the mill* all week or that the demands made on you put you *between a rock and a hard place*. Or even that you're so *dog-tired wild horses couldn't drag you* out of your chair. You're tired and you're willfully relaxing—and that's fine. Put your slippers on and let the cliches do the thinking for you.

But—you wouldn't go to work in slippers. You would hardly conduct business from your easy chair. For your work, and even sometimes for your leisure time, you want your wits sharp and your critical faculties working clearly.

And that's when cliches are bad.

Cliches cloud thought

We reach out for a cliche when we want to make a point and we don't feel like thinking. Sometimes we also use one to show that we're part of a particular group.

If I say, "I'm in a bad place right now," I'm not just telling you that I don't feel good. My cliche also says to you: (a) I want you to realize how very badly I feel—but I can't think of a way to express what exactly I'm feeling; (b) You will understand how I feel because you've used this expression yourself; (c) I'm part of the fashionable circle that uses this up-to-date jargon.

Now, all my reasons for using the cliche may be valid. But—something happened to me when I espoused it. I gave up clarity of thought. I gave up the effort to be specific, to tell you what I felt and why. And I stopped using language as a clear transmission of thought or feeling. Instead, I turned it into a set of stock phrases, each of which is used to cover a multitude of ideas or emotions.

Searching for the words to express exactly what we think or feel can be illuminating, for in the process, we are forced to *define* our thoughts and feelings to ourselves. Our search for the truly apt, specific expression of them gives us a clearer understanding of them. Once we clarify what we're feeling or thinking—we can communicate it.

But reaching for a cliche immediately muddles and muddies the thought. Both speaker and listener start to focus on generalities, not the specific problem or condition that prompted the cliche. And communication becomes a series of vague smoke signals.

In place of cliches . . . straight talk

When you use cliches, you harm yourself. You don't allow language to clarify your thought, and you become a boring, pedantic speaker or writer. I am not suggesting that you try to find an original, colorful expression for everything you think or feel. Wit palls, after a while, as does innovation for innovation's sake. But truth never palls. We *like* to know what each other person is thinking or feeling.

Try to use exact, specific, simple words to express yourself. When you find yourself starting to use a hackneyed expression, stop. Think about it. You may find neither you nor your reader even knows what it means! When you say you feel you've been *dragged through the mill*—what kind of mill are you talking about? Through which part of it were you metaphorically hauled? If you say you had to *pay through the nose*—you are actually referring to an ancient poll tax in Sweden which was levied according to a nose-count. Even if you know that, it has absolutely no relevance for us in America today.

Isn't it absurd to use expressions we don't understand to try to communicate our feelings?

You will find it enormously difficult to escape cliches. We're all so accustomed to speaking in them. They're easy. We don't have to think when we use them. But if you really want to improve your writing, speaking, *and* thinking— you'll get rid of them.

When you feel a cliche coming on, stop, and replace it with an exact expression of your thoughts. For instance, if you should start to say, "He's got us painted into a corner"—stop. Ask yourself *what* he's doing and *why* you feel suddenly at his mercy. Don't let the cliche come in to stop you figuring out what's going on.

Don't try to rid your language of all its cliches all at once. You'd have to stop talking altogether! Start with an hour a day—say, between 9 and 10 in the morning. During that time, determine not to use cliches. First, be aware of them when they pop into your head. Then, try to replace them with an exact transmission of thought.

When this hour's work becomes comfortable, increase it to two hours. And so on.

And remember, it's OK to use cliches when you're taking a rest from thinking. But only when you definitely *don't want* to think. ☐

Some cliches to avoid . . .

we have agreed to disagree
give it the benefit of the doubt
that is the bone of contention
we have built-in safeguards
and by the same token
an exercise in futility
of paramount importance
in this day and age
a few well-chosen words
that's food for thought
from the ridiculous to the sublime
it goes without saying
let's keep our options open
from time immemorial
let's give it the acid test
words fail to express

Enemies of communication

Roadblocks—Part 1

You know the feeling, I'm sure. You're writing along happily, letting your ideas flow, when suddenly — a roadblock appears. You've hit one of those words that momentarily paralyse our thought. A word that seems to be the one we want — or have we got it wrong? Words such as: *like* (or should it be *as*, or *such as*?) . . . *penetrate* (or do I mean *permeate* — or maybe *pervade*?) . . . *abjure* (*adjure*??).

Roadblocks stump all of us. If you are free-writing, as I hope you are (see "The Language of Business," *Tappi J.*, August 1982, p. 117), you jump over the roadblock simply by putting down any of the choices (after all, *you* know what you mean) and allowing your ideas to keep flowing. You can check for the correct word when you come to edit. But, still, the roadblock held you up. It deflected your thoughts.

If you're talking, it's even worse, because you can't go back and edit your speech before others hear it. You're describing your latest project enthusiastically and engagingly to a member of senior management — when you hit one. And say it. And then spend the next five minutes agonizing over the word, instead of focusing on your listener and your subject. Did you choose the wrong word? Did he notice? Why, oh why, did you have to use *that word*?

The big trouble with roadblocks is that they tend to appear in the mind with no conceivable substitute waiting behind them. You can't always avoid them. But you can master the most common ones by listing them, their meaning, and the meaning of the word with which you confuse them. Remove the confusion, and the roadblock's gone.

Let's take on the worst of them.

abjure . . . or adjure?

Here, a touch of Latin can help. The Latin prefix 'ab' means *away from*, as in abort or abduct. So, *abjure* means, literally, *to swear away from*: to renounce, to reject, to abstain from. The Latin prefix 'ad' is the very opposite of 'ab.' It means *toward*. So, *adjure* means *to swear to*: to charge someone to do something, to command him to do it as if under oath.

You abjure smoking. And you adjure your friends to stop offering you cigarettes.

affect . . . or effect?

The rule here is beautifully simple. When you mean *influence: affect* is the verb and *effect* is the noun.

> The president's resignation will *affect* company earnings. It will have a bad *effect* on them.

If you restrict your use of these two words to their meaning of *influence*, they won't give you any more trouble. The noun *affect* is a psychological term that you should save for abstruse discussions on the life of the mind. The verb *effect* is a rather affected (!) way of saying *accomplish* or *make*. "We effected some modifications in our departmental spatial organization" is pompous gobbledygook for: "We changed our floor plan."

all right . . . or alright?

All right. That's all. *Alright* is always all wrong.

any way . . . or anyway?

If you could substitute *anyhow*, you mean *anyway*.

> I have to be in the office by 8:30, *anyway* — so feel free to call me early.

If you mean *any way you like* or *whatever way possible* — write *any way*.

> Get to the office *any way* that suits you — but be there by 8:30.

Pronunciation can help you, here, too. If you find yourself stressing *any*, you mean *anyway*. "I have to do it anyway." If you hear yourself giving equal weight to both *any* and *way*, you need the two words. "Do it *any way* you can."

disinterested . . . or uninterested?

You can be *dis*interested without being *un*interested. *Disinterested* means *uninvolved, neutral*. *Uninterested* means *lacking interest*. A *disinterested* observer, however, may be *interested* in the outcome of a trial.

Reflecting the widespread ignorance of this distinction, some dictionaries have unwisely listed "lacking interest" as a meaning for both words. But if you use *dis-* when you mean *un-*, you'll still displease many an educated ear!

biannual . . . biennial . . . by Gosh!

This is a terrible roadblock that you should really try to avoid. Even if you can remember that *biannual* means *twice a year* and *biennial, every two years* — your readers may think it's the other way around! Determine to say, "every two years" or "twice a year," and you'll all know what you mean.

emigrant . . . or immigrant?

You're an *emigrant* from the country you leave and an *immigrant* to the land to which you move. If you were born in London but live in New York, you're an *emigrant* from England or an *immigrant* to the United States.

factious . . . factitious . . . fictitious . . . or facetious??

With all these words, forget about *facts*. You're not dealing with them.

Factious means *tending to make factions* or to *cause dissension*.

> His remarks that some departments were more important than others caused a *factious* spirit to spread in the company.

Factitious, means *artificial, fake, bogus,* or *contrived.*

> They formulated the lawsuit from *factitious* claims of discrimination, for which no authors or evidence could be found.

Fictitious, on the other hand, relates to *fiction*. It means imaginary.

> Principles may determine the results of *fictitious* trials; personalities have a lot more power in actual ones.

And *facetious* means *witty*, or *attempting wit,* usually through irony.

> I didn't mean you should stop the whole mill for a coffee break; I was just being *facetious*.

Factious looks and sounds rather like *factions,* so that may be a help. The word *factitious* suggests to me *imitation of facts,* which it is. *Fictitious* starts like *fiction*. And *facetious* — well, I think of *faces* — making faces at someone as a rather silly joke.

That's highly personal etymology, but it works for me.

We'll clear away some more roadblocks next time. □

Enemies of communication

Roadblocks—Part 2

Roadblocks are those words that stop us mid-sentence or mid-speech — words which we use fairly frequently but about which we continue to feel unsure. Last time I gave you a list of half the most speech-stumping roadblocks I know. In this column, I'll complete that list for you. However, if you have run into any other roadblocks, just write them down and send them to me. I'll try to clear them up for you in a future column.

flaunt . . . or flout?

Flaunt means to show off, to make an ostentatious display. "The lecturer *flaunted* his expertise by expounding on minute details that were of no interest to anybody."

Flout means to disregard or scoff at something, usually a law or custom. "In his outspoken attacks on other scientists, the lecturer *flouted* conference protocol."

Although there's no etymological connection, it helps me to connect *flaunt* with *flautist* — one who plays his own tune!

flounder . . . or founder?

You *flounder* if you stumble about clumsily or ineffectually. If you completely collapse — you've *foundered.*

Think of a live *flounder* flip-flopping about in a stormy sea vs. the wreck of a ship that *foundered* in the storm.

illegible . . . or unreadable?

A Shakespearean manuscript written in the playwright's scrawling longhand may well be *illegible,* but it certainly won't be *unreadable.* On the other hand, a beautifully typed memo of puffy polysyllabics, cliches, and other business bafflegab is *unreadable,* though not *illegible.*

Illegible refers to the physical act of reading. It means you can't decipher or properly connect the words on the page because the *handwriting* is bad.

Unreadable refers to the mental response you make to a piece of writing. It means you can't take in its meaning because the *style* is bad.

immured . . . or inured?

If you remember that *murus* means *wall* in Latin — or that *mur* means *wall* in French — you won't have a roadblock here. *Immured* means, literally, *walled in*— imprisoned, entombed. *Inured* means *resigned to by habit.*

You can be *inured* to the practice of making a quarterly budget and *immured* in paperwork while you're doing it.

principle . . . or principal?

Despite the nationwide thunderings of stylists and English teachers, these two continue to be confused. A *principle* is a deep belief; a *principal* is the headmaster or headmistress of a school. Now, the adjective "principal" means most important, consequential, or influential. That's easy to remember because, as kids, we certainly considered the *principal* the most important, consequential, or at least influential figure in the school.

And what about the adjective, *principle?* It doesn't exist. *Principle* exists ony as a noun.

Think of a school *principal,* and you'll remember which word means important, consequential, or influential. It works as both a noun and an adjective.

seasonable . . . or seasonal?

If you talk about a *seasonable* rise in consumer spending, you mean it came just at the right time. But if you call it *seasonal,* you are saying it was determined by or related to that particular season.

Seasonable has come to mean timely, opportune, suitable to the concomitant circumstances. *Seasonal* means simply "of the season."

stolid . . . or solid?

If you call a person *stolid,* you're saying he expresses no emotion or sensibility. He's impassive.

If you say he's *solid* — you mean he's whole, real, trustworthy, straightforward, dependable.

Someone can be *solid* — a positive trait — without being *stolid* — a term with a slightly negative cast.

"Solid facts" is a redundancy; facts are either real and complete or they're not facts. "Stolid facts" is redundancy bordering on the ridiculous, since no fact could possibly express an emotion, however hard it tried!

strategem . . . or strategy?

A *strategem* is a cleverly contrived trick performed to gain a particular prize or to deceive the other party. You should not talk about your company's "strategem for growth" unless you mean to imply that the firm has a nasty trick or two up its collective sleeve.

A *strategy* is simply a skillful, detailed plan to reach a goal. It can be perfectly decent, open, and harmless. It's usually the one you mean.

tortuous . . . or torturous?

People almost always mean *tortuous* when they use either of these words. However bad things in business may be, they can rarely be described as *causing torture* or *cruelly painful* (the definition of *torturous*).

Tortuous comes directly from the Latin *tortus*, or twist. It means twisting, winding, or crooked.

Torturous also comes from *tortus*, but indirectly — via a more *tortuous* path. . . . It's the adjective for *torture*, which of course means to twist or distort with pain. Remember that *torturous* has two "r's" like its noun, *torture*, and you won't confuse the two any more.

turbid . . . or turgid?

The final roadblock is easy to clear away. *Turbid* means muddy, clouded, obscured in a mass of confusion. Think of the cloud of exhaust released by a *turbojet*. Then you'll be able to distinguish *turbid* from *turgid*, which means inflated, swollen, or bombastic.

Turgid is usually used to refer to the kind of pompous language that this column is trying to eliminate!☐

Snarks and bojums

"Then you should say what you mean," the March Hare went on.

"I do," Alice hastily replied; "at least—at least I mean what I say—that's the same thing, you know."

"Not the same thing a bit!" said the Hatter. "Why, you might as well say that 'I see what I eat' is the same thing as 'I eat what I see.' "

—Lewis Carroll, *Alice's Adventures in Wonderland.*

The Hatter's right, you know. We may *think* we mean what we say. But, all too often, we don't! All too often, our utterances reflect the whimsies of Wonderland, and our readers are treated to the kind of baffling "explanation" that makes sense only to the writer. An explanation like:

"For the Snark *was* a Bojum, you see."

—Lewis Carroll, *The Hunting of the Snark*

Sure we see!

The offspring of "Snarks" and "Bojums" have wriggled their wily way into business language. They are that happy family of nonsense words—words that don't appear in the dictionary in the way the speaker or writer uses them. These words and expressions mean either nothing at all (misspelled words or mixed metaphors that don't make any sense) or something entirely different from the meaning the speaker assumes for them. Let's call the first type "snarks," and the second, "bojums."

Snarks are not confusing; they're just wrong. And they can be annoying and distracting. Recently, I read a paper in which the writer kept hoping he wouldn't "embarass" his colleagues. Probably the word occurred no more than four or five times, but I found my irritation growing with each appearance, until at last it killed my interest in the paper. My reaction was perhaps extreme; nevertheless, I am told that it is not uncommon.

The fact is, snarks can deflect your message. A bit of attention to the number of r's and s's and the order of i and e can do wonders for the effectiveness of your writing! Here is a list of words whose spelling bothers all of us:

accommodate — 2 c's and 2 m's
characteristic — first a, then e
embarrass — 2 r's and 2 s's
harass — 1 r, 2 s's
judgment — no e between g and m
liaison — the a is sandwiched between 2 i's
liquefy — with an e
minuscule — with a u
prevalent — a, then e
sacrilegious — i, then e
seize — e before i (remember: you have to see to seize!)
supersede — all s's, no c's
tranquillity — 2 l's
vilify — 1 l
weird — e before i

Instead of trying to form elaborate mnemonic games (which you may forget!), you'll find it easier just to keep this list of words available for consultation. Soon, you won't need it.

The other sort of snark defies lists, for you can mix *any* metaphor that you use without thought. Again, mixed metaphors rarely confuse; they just annoy. When you tell your staff: "you've got to keep your shoulder to the grindstone and your nose to the wheel," no one is likely to mistake your message. But the image you invoke will get some discreet snickers and rob your words of power. Here are some mixed or senseless expressions that have appeared in memos and papers. You can avoid these and others by thinking clearly and not reaching for a cliché when you feel too lazy to say what you mean.

"It's a mute point."

"He threw a cold shoulder on *that* kettle of fish."

"You must get them to tow the line."

And, in a Washington *Post* article quoting Edward E. David, Jr., President Nixon's last science adviser:

"The White House advisers to Mr. Nixon thought that the scientists were using science as a sledgehammer to grind their political axes."

Bojums add confusion to annoyance, for they actually rob the word or expression of its meaning, and so mystify the reader. The most common bojums are words mistaken for others they resemble:

ascent — n. rise
assent — n. agreement

biannual — twice a year
biennial — once every 2 years or lasting 2 years

climactic — relating to a climax
climatic — relating to the climate

corespondent — one charged as the respondent's paramour in a divorce suit
correspondent — a letter writer

complement — something needed to make a thing complete
compliment — a word of praise

immured — walled in
inured — accustomed

ingenious — inventive
ingenuous — naive

These are only a few of the problematic pairs; we'll look at others another time. Let's consider some of the most infamous bojums, whose misuse is due not to confusion with a near-twin but to sheer lack of understanding:

fortuitous — what happens by chance, *not* fortunate

gratuitous — unearned, unmerited, *not* free

fulsome — excessive, disgusting, *not* generous

hopefully — full of hope, in a hopeful manner, *not* "it is to be hoped"

unique — without equal (so you can't have *more* or *less* unique)

prototype — primal or original type, *not* a representative or example

Snarks *are* bojums, in that both make nonsense out of language. Keep these basic lists as a memo to yourself for, to return to Lewis Carroll's wisdom:

. . . the King went on, "I shall never, *never* forget!"

"You will, though," the Queen said, "if you don't make a memorandum of it."

A checklist of problem words

Lately in my writing courses, I have noticed a regular phenomenon. People who are doing beautifully on a rather difficult exercise—such as performing a needs analysis or free-writing for ten whole minutes—suddenly look up with an expression of acute pain or anxiety.

"Is it **affect** or **effect**?" they ask fretfully. Or, "How can I say 'The main criteria for judgment **are** . . .' when there's only one criteria??"

Again and again the same problem words come up to stall writers in their efforts to achieve a free-flowing, readable piece of work. I have noted the ones most frequently mentioned in my seminars or misused in students' writing—and I offer them here as a checklist for you to use whenever they pop into your mind to trouble you.

Cheryl Reimold is president of PERC Communications, 6A Dickel Rd., Scarsdale, N.Y. 10583.

To make this a quick and easy reference source, I have capitalized the word you usually want. If the word has several meanings, I have given only those that you are likely to need. I have included a memory-helping tip whenever possible. And, in case the word persists in perplexing you, I've suggested some alternatives!

First, let me give you two general rules that will help you deal with all problem words:

1. Keep a copy of a good pocket dictionary on your desk.
2. If you're unsure about a certain word, use another one.

For instance, if you can't remember whether you mean **affect** or **effect**, and you don't have this list or a dictionary nearby, you could just say **influence.**

Now let's look at the words themselves (Table I).

I. Problem words checklist

TO ADAPT . . . to change to fit a new use or new conditions
TIP: Remember that **apt** means "suitable" or "fit."
to adopt . . . to accept, to take on
TIP: You don't **adapt** a child, you **adopt** one.
"We would like to **adopt** your plan for reorganization, provided you allow us to **adapt** some of its components to the particular needs of this department."
ALTERNATIVES: **adapt** = "change to fit"
adopt = "take up"

TO AFFECT, verb = to influence
EFFECT, noun = result
EFFECTIVE = that which causes the desired result.
TIP: These three are the ones you want: the verb **to affect**, the noun **effect**, and the adjective **effective**.
to effect, verb = to cause
the affect, noun = a highly technical psychological term you'll never need in your business writing
affective = relating to, a rising from, or influencing feelings
TIP: You never need these three in business writing.
"This method of coating is more **effective** than I had supposed. I did not expect the method to **affect** the brightness of the sheet. Indeed, the **effect** of adding certain chemicals later in the process appears to be an increase in brightness and a decrease in density.
ALTERNATIVES: **to affect** = "to influence"
effect = "result"
effective = "useful"

council, noun = a group or assembly
COUNSEL, noun = advice
TO COUNSEL, verb = to give advice
"I would **counsel** you to listen to all sides carefully before you decide. Don't accept anyone's **counsel** without considering the consequences of taking the suggested action."
ALTERNATIVES: **counsel** = "advice"
to counsel = "to suggest"

CRITERIA = standards or characterizing traits, **always plural**. If there's only one, it's a **criterion**.
"The **criteria** for success in this experiment are high brightness, low density, and low cost. Good recovery of chemicals is not a **criterion** at this stage."
ALTERNATIVE: "standards" or "gauges"

DATA = facts, **plural**. The singular, **datum**, is rarely if ever used.
"I have almost all the **data** on the experiment; only one small result remains unexplained."
ALTERNATIVE: "information"

FORTUITOUS = accidental, **not** lucky
TIP: Don't use this word! If you mean lucky, say "fortunate." If you mean "a lucky accident," say so.
"The discovery of penicillin was a lucky accident."

immanent = dwelling within
IMMINENT = about to happen, in the offing
TIP: think **reMAIN** for **immANent**; **IN the offing** for **immINent**
"Labor leaders said a strike settlement was **imminent**."
ALTERNATIVE: **imminent** = "about to occur"

TO MILITATE (AGAINST) = to operate against
TIP: think of **militia** or **military**
to mitigate = to make less harsh or severe
"The results of his own previous tests militated against his new theory."
ALTERNATIVES: **to militate against** = "to work against"
to mitigate = "to alleviate"

I hope this checklist militates against any imminent problems in your writing! Good luck, and keep the list handy.

In small scale

Most writers—particularly technical writers—know the big words well enough. Few scientists or engineers would confuse "explosion" with "implosion" or even "hydroscope" with "hygroscope." But, throw a bunch of little prepositions into the grab bag—and what we come up with is anyone's guess! "Forbid *from*" or "forbid *to*"? "Adapt *for*," "adapt *to*," or "adapt *from*"?

It can be headachy business. And—the niggling doubt persists—do those annoying little words really matter?

Yes, and not only because the English language is worthy of care. Small as they are, these words can cause abrupt, unwanted shifts in meaning if they are misused.

Take "adapt." "Adapt *for*" means "prepare for a specific purpose": The novel was *adapted for* the stage. "Adapt *from*" means "take or abstract from something else": The painting was *adapted from* a photograph. "Adapt *to*" means "adjust to fit certain conditions": The species was able to *adapt* itself *to* the new climate.

Or, consider "concerned *about*" vs. "concerned *in*." If you are *concerned about* the energy crisis, you're worried about it. If you're *concerned in* it, you're a part of it!

Little words can mean a lot. If you use the wrong one, it can throw your readers off balance, or at least annoy them. Neither will help get your point across. Remember which prepositions to use and when—and you'll be able to spend your time on the content of your writing instead of worrying about whether to use "at," "to," or "for"!

Here's a list of some common confusions in preposition-land.

admit *to* = to confess

Jones *admitted to* falsifying the documents.

admit *into* = to allow entry

My colleague had us *admitted into* the club.

agree *to* = to give assent

The director *agreed to* the change in schedule.

agree *with* = to be of the same opinion

I *agree with* you; something must be done immediately.

agree *on* = to be in accord with someone about something

At last we *agreed on* a plan.

anxious *about* = worried about

We are all *anxious about* Allen; he is still out sick.

anxious *to* = feeling pressed to

Now that the results are in, we are *anxious to* move on to Step 2.

compare *to* = to point to likenesses

The speaker *compared* our methods *to* those of our competitors.
(He pointed out the similarities between ours and theirs.)

compare *with* = to point to differences

The speaker *compared* our methods *with* those of our competitors.
(He pointed out the differences between them.)

correspond *to* = fit, match

The model does not *correspond to* actuality.

correspond *with* = exchange letters

J. S. *corresponds with* us fairly regularly.

differ *from* = to be different from

He *differs from* me in personality, not beliefs.

differ *with* = to disagree with

differ *on* = to be in disagreement about something

I *differ with* you *on* the source of the problem.

impatient *at* = refers to behavior

I was *impatient at* his lack of control during the meeting.

impatient *with* = refers to a person

Don't be *impatient with* trainees who are trying to learn.

in behalf of = in someone's interest

I felt I had to argue *in his behalf.*

on behalf of = representing someone

I am appearing *on behalf of* Dr. Wallace.

Now, here are some accepted prepositional idioms vs. unacceptable ones. If you use the wrong one, you probably won't confuse your readers, but you may hold them up and stop their flow of comprehension as they try to figure out "what jars here."

ability *to* (**not** *of*)

You have the *ability to* do far more than you are doing now.

according *to* (**not** *with*)

According to Mr. Simons, the profit picture is looking brighter.

acquiesce *in* (**not** *to*)

The committee voluntarily *acqueisced in* the chairman's proposal.

apropos *of* (always followed by *of*)

Apropos of your suggestion, we plan to adopt the new techniques next time.

authority *on* (**not** *about*)

He is a world *authority on* solar energy.

comply *with* (**not** *to*)

If they do not *comply with* the new regulations, they will have a stiff price to pay.

forbid *to* (**not** *from*) but prohibit *from* (**not** *to*)

He said he would *forbid* them *to* leave (*prohibit* them *from* leaving) before the session was over.

identical *with* (**not** *to*)

These samples are *identical with* those you sent us last year.

oblivious *of* (**not** *to*)

The President's mother was clearly *oblivious of* the sensation caused by her remarks.

Finally, a preposition's disappearance won't pass unnoticed. If you use two words that require different prepositions, remember to include both prepositions.

The visual aids increased our interest *in* and attention *to* the talk. (*Not:* . . . increased our interest and attention to the talk.) The foreman said he was afraid *of*, but not deterred *by*, the wires in his path. (*Not:* . . . afraid, but not deterred by, the wires. . . .)

If both words take the same preposition, you need use it only once.

He said he could comply, though not agree, with her demands.

Are you worried about ending a sentence with a preposition? Don't be. Sometimes you have to! Remember Churchill's retort when he was accused of this grammatical "crime":

"This is the sort of thing up with which I will not put."

Nouns and nounery

Nouns are the "heavies" of language. They plant themselves firmly in sentences and phrases, and there they sit, unbudging, immovable. Some style-conscious writers have bemoaned the "noun-demons" that ruin language, but the term is not really apt. Demons move. Nouns don't.

Think about the reports, memos, or letters that you write. Most of the time, you're (a) describing an event or a process, (b) explaining how to do something, (c) asking for something to be done, or (d) putting forth a concept or theory. You're leading from one thought or action to the next, and you want language that *moves*, easily and freely, to translate the flow of thought into words. Words that stop the flow of a phrase confuse or hobble the thought. The "hobblers" are almost always nouns.

A scourge of technical language

To free your language, begin by eliminating *nounery*: the overuse of nouns. The main types of nounery are:

- Noun clusters: "cost impact considerations" (price); "schedule stretchout adjustment" (delay) . . .
- Pompous words with fuzzy meanings: "of the order of magnitude" (about); "paradigm" (model); "in view of the foregoing" (so) . . .
- A preponderance of prepositional phrases: "as a result of" (because); "in excess of" (more than); "in the amount of" (for) . . .

Technical writing is emblazoned with nounery. A research director may tell his assistant, "You're not to change anything before you've checked it thoroughly," but put a piece of paper in front of him or her and you're likely to find:

> All aspects of the situation should be taken into careful consideration prior to the implementation of any corrective action.

Yes, it sounds absurd—but take a look at the recent memos and instruction sheets circulating in your department!

Now, there are three main causes of nounery in technical writing:

- A desire to lend weight to one's words. The research director feels that instructions concerning the "implementation of corrective action" will be taken more seriously than words about "stopping the leak." In fact, the latter is clearer and has more chance of being understood.
- The scientist's preoccupation with *naming things*. Science names the world, every bit of it, and in so doing makes it a bit more understandable, or at least more acceptable, for the rest of us. But—in ordinary interdepartmental memos, it is not necessary to name and describe every inessential facet of the subject. Indiscriminate naming and describing rarely illuminate your subject or your reader; they confuse both.
- Paradoxically, the desire to be short and concise. This holds true particularly for the notorious noun clusters—groups of nouns in which all but the last are forced to act as adjectives. "Bridge structure modifications" contains only three words, but "changes to the structure of the bridge" is much easier to read and comprehend. And effective writing is that which is understood quickly—not that which uses the fewest words!

Once you recognize the causes and symptoms of nounery, it's fairly easy to avoid it. Are you assiduously naming things that your reader doesn't really need to understand? Are you trying to impress? Have you formed a tangled clutch of nouns that could be freed into a flow of adjectives, adverbs, and, especially, verbs? "Make use of company expertise in marketplace knowledge" sticks. "Use your firm's knowledge of the marketplace" flows. "Subcontract modifications frequently give rise to increased expenditures" is a muddled mouthful. "Changes in the subcontract can increase costs" is not.

Commonly abused nouns

Nounery is not the only nominal obstacle to clear writing. There are also the words themselves—many of which are not only overused but misused. Some of the most commonly abused nouns are:

Affect, Effect. The rife confusion between the verbs *affect* and *effect* has affected the use of the nouns! *Effect* is almost always the noun you want; it means the result or consequence of an action. *Affect* has a very limited psychological meaning, defined by Webster as "the conscious subjective aspect of an emotion considered apart from bodily changes."

Dilemma. This term means a choice between *two* disagreeable alternatives. Anything else is a *problem*, a *predicament*, a *plight*, or a *difficulty*.

Method, Methodology. *Methodology* is the *study* of methods. Chances are you're talking about the *method*—a systematic procedure or technique—not its analysis.

Allusion, Reference. The misuse of these words is abetted by spelling confusions: *illusion* (a false image) and *elusion* (an adroit escape) appear all too often in the place of *allusion*. An *allusion* is an indirect mention; a *reference* points directly to the subject. "As Einstein pointed out . . ." is a reference to the great physicist; "$e = mc^2$" is an allusion to his theory.

Reversal, Reversion. Think: *reversal—reverse; reversion—revert*. A *reversal* is a turnabout—"the 2-day strike reflected a reversal in union policy." A *reversion* is a return to a former state, as "a reversion to primitive behavior."

If you free your language of nounery and seek to give words their explicit meaning, you will find that your reports are easier both to write and to read. Instead of the following (actual) monstrosity:

> Of extreme importance in timely and effective processing of changes to major subcontracts with a minimum of adverse cost impact is the maintenance of effective upsteam visibility by the subcontract administrator.

your memorandum would read simply:

> To keep costs down, the subcontract administrator must inform top executives of any changes made to the original contract.

Half the number of nouns—twice the clarity!

Being verbal—bringing life back to verbs

We live, we work, we think, we act, we meet, and, sometimes, we disagree. Translate that into business language, and you get: We are alive, we are employed, we take things into consideration, we take action on them, we gather together, and, on occasion, we have disagreements. Can you feel the heaviness, the dullness—the *death* of the living activity those words were supposed to convey?

Simple, active verbs breathe life into writing; multi-word phrases stifle any quiver that might have been there. Verbs are the most ignored, most effective tool of good business and technical writing. Yours will grab your reader's attention and transmit your message with brilliant clarity *if* you follow a few simple guidelines.

1. Write what you think. Superloquacious pomposity seems to pour mainly from the pen; we actually think in simple, active verbs. If you think: "We must keep our program reports up to date," write it. Don't fill up your memo, as one executive did, with: "Prompt and effective monitoring of the program reports is an absolute necessity to assist in assessing performance data..."

2. *Avoid* the passive voice. Justly condemned as the most deadly of technical writers' paradigms, the passive voice is singularly boring, uninformative, and lifeless. It is also ubiquitous. Judging from reports and memos, no recognizable individual ever *does* anything in business; activities are inscrutably performed by some unknown, or unnamed, power: "It is recommended that a fan pump be installed on No. 9 paper machine..." Who recommended it? Who will install it? "A wide variety of composite subjects were discussed..." By whom?

Occasionally, we do learn the identity of the agent: "The headbox consistency is fixed at 0.92 by drainage considerations." Why not: "Drainage considerations fix the headbox consistency..."?

Now, what's wrong with the passive voice? First, it slows down the sentence with extra words. It also slows down the reader. In life, people or events *cause things to happen.* A boy sets a fire, for instance. If you write, "That fire was started by a teenage boy," the reader mentally translates it into the logical, actual sequence: "A teenage boy started the fire." That extra translation takes time and stops the flow of writer–reader communication. Furthermore, the word order of the passive voice actually reverses the logical progression of the action; instead of *boy—start—fire*, we get *fire—start—boy*. We don't follow the action, our minds don't form images—the sentence dies.

The passive voice can also be terribly confusing. It engenders all sorts of misplaced modifiers. In a book recently submitted for publication, we find: "Assure that a program manager is established by the subcontractor with authority to speak for and commit his company..." What the writer meant was "Assure that the subcontractor *appoints* a program manager with authority to speak for and commit his company..."

Even in a grammatically correct sentence, the passive voice can cause momentary disorientation, as it *distorts* the natural order of subject—verb—object. We lose track of who did what to whom. "Bill thanked Joe for his help" is immediately clear. "Joe was thanked for his help by Bill" is not.

So—eschew passives whenever you can.

3. State positive facts *positively.* To avoid taking a stand or committing themselves, business writers often resort to double negatives. "General dissatisfaction *cannot but* lower productivity" is a jumbled mouthful. "General dissatisfaction *can only* lower productivity is clearer.

Similarly, "The profits were *not less than 50% in excess of* projected figures" can be rendered comprehensible by writing simply: "The profits were *at least 50% higher* than projected figures."

4. Don't misuse or abuse your verbs. Some verbs seem to lend themsleves to constant misinterpretation. Others attract a following of uncessary prepositions and nouns that serves only to puff up the sentence and smother the verb. You'll write the right verb if you remember that:

- *Abjure* means to renounce: "He *abjured* his oath of secrecy."
- *Adjure* means to require or charge: "The judge *adjured* him to speak, despite his oath of secrecy."
- *Adapt* means to adjust or make suitable: "We had to *adapt* ourselves to the new schedule."
- *Adopt* means to accept or receive: "We decided to *adopt* the new schedule."
- *Comprise* means to contain: "The main office *comprised* 10 divisions."
- *Compose* and *constitute* mean to make up a whole; "Together, the divisions *compose/constitute* the main office."
- *Consist of* means to be made up of: "The office *consists of* 10 divisions."
- *Transpire* means to become known, *not* to come to pass: "From the talks it *transpired* that the president was in favor of expanding research."

Then—look at your verbs. Can you rescue any from useless hangers-on?

"He appears (*to be*) unhappy with the decision." Appears means "seems *to be.*"

"She was appointed (*to the post of*) treasurer." You are always appointed *to a post.*

"Jim commutes to (*and from*) Washington." You can't commute one way.

"The project continues (*to remain*) on schedule." If it continues, it automatically remains.

"These problems date (*back*) to the last administration." How could they date forward?

Remember—you're trying to communicate complex technical information *clearly, quickly,* and *appealingly.* Verbs can do it for you.

How to put power into your language

Part 1: Use one-step words

Releasing the power of language

I hope the series on poetry helped illuminate some facets of the power inherent in language. We can begin to release this power by *choosing* and *combining* the elements of language not haphazardly but by design—as poets do.

In this series, we'll look at three elements of language: words, sentences, and paragraphs. We'll see how to make good choices of each and then how best to combine the choices.

Let's begin with the choice of words. In *English Prose Style*, Herbert Read writes:

> an isolated word . . . must *mean* the thing it stands for, not only in the logical sense of accurately corresponding to the intention of the writer but also in the visual sense of conjuring up a reflection of the thing in its completest reality.

Here, I think, is a near-perfect definition of *the right word*: one that can "conjure up the reflection of the thing in its completest reality." The right word is always a "one-step word."

One-step and two-step words

I call "one-step words" those that are *one step away* from the object or concept itself. A one-step word is the closest to the object; no other word stands between it and the thing it denotes.

"Two-step words" stand two steps away from the object they depict. A two-step word does *not* immediately "conjure up a reflection of the thing." Instead, it conjures up the one-step word that stands between it and the thing it denotes.

If I say, "I saw a cat with an egg-shaped head," a rather bizarre furry animal will immediately appear on your mental screen. But if I say, " I perceived a feline animal with an ovate cranium," no cat will leap into your mind. Rather, you will first find yourself replacing the two-step words (*perceived, feline, ovate,* and *cranium*) with their one-step equivalents (*saw, cat, egg-shaped,* and *head*). Only then will your cat come into view!

Here are some one-step words with their two-step equivalents: *door* (portal), *house* (domicile), *money* (financial affairs), *mad* (irate), *big* (sizable), *walk* (perambulate), *begin* (initiate), *end* (terminate).

As you can see, one-step words for everyday things tend to be short, Anglo-Saxon words. Indeed, many writing teachers tell their students to stick firmly to the "rugged" Anglo-Saxon words, avoiding "fancy" Latinate words whenever possible. This advice is misleading, however, since we don't limit our conversation to cats, houses, and walks!

Look at these one-step words for more complicated objects or events: *calendar, computer, confused, delegate, negotiate, television.* Each word is the closest to the object or event it denotes. And each is fairly long, with a Latin or Greek root.

We are not trying to choose short, simple words but *one-step words*—words that conjure up the reflection of the thing meant.

Choosing one-step words

If you can't tell a one-step from a two-step word by its length or linguistic heritage—how do you find the words you want?

You find them by *analyzing the effect of the words you read.* Try to do this exercise daily for the next two weeks. Circle two paragraphs: one from a business letter or memo, the other from a good book, poem, or newspaper. Read each paragraph in its entirety. Then go through it word by word to see what the words show you. Do they immediately conjure up reflections of objects and events (one-step words)—or do they make you search your brain for words closer to the thing meant (two-step words)?

For example, consider this sentence from Paul Sheehan's article, "A Preserve for Wildlife and Art," in *The New York Times*, Sunday, February 23, 1986: "A visitor feels the exhilarating chill of meeting the yellow gaze of a crocodile just 20 feet away."

All the words are one-step; we stare right at that crocodile. Now here's a two-step translation, with apologies to Mr. Sheehan: "A sojourner in the area becomes aware of a tingling sensation, gelid but not disagreeable, as he makes visual contact with the ochre-colored eyes of a nearby crocodile."

As you read to see what the words are showing you, three things will happen. First, you'll savor the good writing. Second, you'll find you're substituting one-step for two-step words in the bad. And third—you'll start using one-step words!

How to put power into your language

Part 2: Write sentences that work

The simplest sentence transmits one of three messages:

- What happened (is happening, will happen) . . .
  "Spot runs." "I fell."
- Who did (is doing, will do) what . . .
  "Spot chased the cat." "I will take the train."
- How something is (was, will be). . .
  "Spot is tired." "I will be late."

Everything else—all the adjectives and adverbs and phrases and clauses—are simply elaborations on one of these themes.

Now, an effective sentence is one that gets its message across! It does so by being clear, understandable, and interesting. If it's not interesting, it won't be read.

If you adhere to the following three rules, your sentences will work for you.

Rule 1: One thought to a sentence.

The rule is clear. But the question is: How do you decide what constitutes a single thought?

Consider this sentence:

> The meeting, which was scheduled for May 23 but had to be cancelled because several of the directors could not be present on that date, has now been rescheduled for June 9, provided there are no conflicts with any of the participants' plans for that day.

Despite the many clauses, the writer of this cumbersome sentence could argue that he expressed only one thought. And, indeed, his sentence demonstrates that an ordinary thought can be complex and multidimensional. You can go on and on adding ramifications and qualifications, making your sentence more and more unreadable with each one! How do you know when to stop a thought and continue it in another sentence?

Here's a practical answer. Limit each sentence to two, or at the most three, clauses. The easiest way to do this is to count the verbs in your sentence. If you find more than three, try to break the complex thought into two or more simpler ones, with a sentence for each. The example above would then be:

> The meeting, originally scheduled for May 23, had to be cancelled as several directors could not be present then. We plan to hold it on June 9, if participants have no conflicts for that date.

Rule 2: Vary the length and structure of your sentences.

This rule also takes work. It's much easier to write all your sentences the same way:

> We hope you will be able to attend the meeting. We think you will find the program varied and interesting. We request that you inform us of your decision by May 15.

These three sentences will not inspire the reader to come to your meeting. To make him feel the variety and interest of your meeting, you must put variety and interest into your sentences:

> We hope you will be able to attend the meeting. The program we've planned is varied. It's interesting. And it will certainly be enhanced by your contributions. If you can let us know your decision by May 15, we'll be most grateful.

Determine to vary the structure of your sentences. If you started the last one with a pronoun and a verb ("We hope. . ."), start this one differently ("The program we've planned is. . .").

Then look at the length of the last sentence. You don't have to go from long to short and back again. That would look artificial and silly. But try to vary the word count slightly, and every so often, put in a short sentence ("It's interesting.").

Rule 3: Emphasize important words by end placement or inverted word order.

This rule, while easy to adopt, goes contrary to what most of us do. We tend to put the most important part of our message first—because it is foremost on our minds. We write:

> Productivity is the most important measure of our success.

More effective would be:

> The most important measure of our success is productivity.

For the sake of variety, you will not always want to place the most important words at the end of your sentences. Another way to emphasize them is to use inverted word order. Suppose you want to point to a certain type of paper you use for bags. This sentence won't do it:

> We use this paper for heavy-duty bags because of its toughness.

Instead, you could invert the normal word order and write:

> This paper we use for heavy-duty bags, because of its toughness.

Try these rules, one at a time if you like—and make your sentences work for you!

How to put power into your language

Part 3: The paragraph = A unit of thought

Here's a law of paragraphing:

> Readers abhor a lack of paragraphs.

They like pages with several paragraphs of varying length. If you vary your writing with paragraphs, people will want to read it. If you don't, they won't. It's that simple!

Frequent paragraphing is so important to readers because it fills three of their basic needs:

1. A sequential, logical development of the message—one thought or step at a time.
2. Breathing space between ideas or steps.
3. Visual variety.

A page of writing without paragraphs is too much to take in all at once. It does not enable the reader to see where one thought or step ends and another begins. It denies him the necessary break to pause and digest an idea before moving on to the next one. And it offends his eye. Remember, your reader *sees* your page of writing before he reads it. If it looks offputting—as a dense, unbroken mass of type inevitably does—he will not want to read it. You will predispose him against it. Technical writers may be shaking their heads and mumbling, "Doesn't apply to technical material." On the contrary. Readers of difficult technical material often need even more guidance from the writer.

To make logical, helpful paragraphs, take your freshly written draft and separate it into *units of thought or action.* Start a new paragraph at the beginning of each new division. If you find a paragraph running longer than 10–12 lines, try to divide it further or look for another way to express your idea. Here is a single paragraph that appeared in *European Science Notes.**

> It is well known that polymer additions can cause reductions in pressure drops (and therefore pumping power) on the order of 40 to 60 percent. What is less well known is the mechanism responsible for this. This requires measurement of the turbulence quantities in the near-wall region. Special problems arise in attempting to make LDV measurements near walls. As was pointed out by F. Durst (University of Erlangen-Nurnberg), this requires, first of all, the use of a probe volume that is properly suited for the wall region. He suggested a probe diameter of less than 60 μm in diameter. It is worth pointing out that most commercial systems produce probe volumes that are anywhere from 100 to 200 μm in diameter. In addition, sufficient sampling time must be provided if smooth measurements are to be abtained. Attention to the integral scales indicates that for a suitable number of samples, say 1000, 100 seconds per point are required. Finally, because of the curvature of the inner and outer walls, special problems arise in attempting to make measurements (even away from the wall) in the circular glass pipes which are used for these studies. This is because of the refraction of the beam which occurs as it reaches the fluid within the pipe and is transmitted back through the glass and the air on its way to the receiving optics. In such cases it is difficult to know where the probe volume is located. In addition, such large beam reflections may be produced that it is geometrically impossible to carry out the measurements.

To make this paragraph more readable, we can divide it into the following new paragraphs, each one a *thought-* or *action-unit.*

> It is well known that polymer additions can cause reductions in pressure drops (and therefore pumping power) on the order of 40 to 60 percent. What is less well known is the mechanism responsible for this. This requires measurement of the turbulence quantities in the near-wall region.
>
> Special problems arise in attempting to make LDV measurements near walls. As was pointed out by F. Durst (University of Erlangen-Nurnberg), this requires, first of all, the use of a probe volume that is properly suited for the wall region. He suggested a probe diameter of less than 60 μm in diameter. It is worth pointing out that most commercial systems produce probe volumes that are anywhere from 100 to 200 μm in diameter.
>
> In addition, sufficient sampling time must be provided if smooth measurements are to be obtained. Attention to the integral scales indicates that for a suitable number of samples, say 1000, 100 seconds per point are required.
>
> Finally, because of the curvature of the inner and outer walls, special problems arise in attempting to make measurements (even away from the wall) in the circular glass pipes which are used for these studies. This is because of the refraction of the beam which occurs as it reaches the fluid within the pipe and is transmitted back through the glass and the air on its way to the receiving optics. In such cases it is difficult to know where the probe volume is located. In addition, such large beam reflections may be produced that it is geometrically impossible to carry out the measurements.

The first paragraph is an introduction to the topic. It states that the poorly understood mechanism requires measurement of turbulence near the walls. Each of the next three explains a specific difficulty inherent in this type of measurement.

The newly paragraphed version is easier on both the eye and the mind, as it provides both visual variety and a structural guide to the different topics discussed.

From now on, check your writing for *thought-units* or *action-units*—and give each a paragraph. Next time, we'll see how to construct each paragraph for maximum clarity and power.

Cheryl Reimold is president of PERC Communications, 6A Dickel Rd., Scarsdale, N.Y. 10583.

*Brown, Eugene F., "The First International Conference on Laser Anemometry," *European Science Notes*, Vol. 40, No. 5, May 1986.

How to put power into your language

Part 4: How to produce a powerful paragraph

In the last column, I defined a paragraph as a **unit of thought** or a **unit of action**. To make your paragraph a powerful part of a tight, strong piece of writing, you must make each sentence serve that focal thought or action.

Find the unifying thought or action

To find the unifying thought or action, read your original paragraph, close your eyes, and ask yourself: What is this about? When you find an answer, see that every sentence relates directly to that central idea.

If you can't find a single thought or action, or if you find more than one, your paragraph lacks the necessary unifying force. You will need to establish a single idea that imbues every sentence of the paragraph. Here are some ways to do this.

Make a statement and amplify it

Write down a core statement of your message. Then make a list of connective words: **and, but, then, however, because, that means, therefore, for example**, and others that occur to you. Now choose one of these words that leads you to a follow-up statement, and write the statement. Go on doing this for a few more sentences. For example:

> We need more communication in our department. AND it has to come from the top. BECAUSE the people running a department set the standards for the way to behave. THAT MEANS we must persuade our department leaders to endorse a policy of more open communication. THEN, they might organize an informal department meeting to discuss ways of communicating better here.

Obviously, the final paragraph would not consist of sentences all beginning with a connective word. You would improve the style. But this exercise will help you find and develop a unifying thought—in this case, the need for leaders to endorse and exemplify a spirit of communication.

Give questions and answers

This mode of organization employs a little suspense. Write your central thought as a question and then suggest answers:

> What should department leaders try to accomplish with this staff meeting? Could it be more than a policy-setting session? Could the department leaders, in fact, use the meeting as their first opportunity to communicate better with the staff?

The first question is the only real question here; the other two are actually answers dressed up as questions. By putting your message in the form of three questions, you stimulate the reader to pause and think out the answer himself, before going on to the next sentence. This style involves the reader and encourages him to think through the issue himself.

Describe a process step-by-step

This is often the simplest and best approach when you want to tell the reader **how something works** or **what happened**. For instance:

> You push the pencil through the hole in the electric sharpener. As soon as the tip touches the blades, they start grinding it. When it is honed to a fine point, they automatically stop. You then remove the pencil.

Of course, a whole report written like this is deadly to read—as indeed is any piece written in any one style. Good writing demands variety. The next paragraph should take up some phase of the operation that you can discuss in a different way.

Imply the unifying thought

If you want to communicate a feeling, the most effective way is usually to lead the reader to get the feeling from the facts or events you relate. Don't state the feeling itself. Here's an example:

> The only street light shone on the backs of a couple walking away from the station. The whistle of the last train had peaked and trailed off five minutes ago. He looked at his watch automatically, for something to do, and mumbled, "9:05." No one heard him. Even the evening mist which had clung to his face and hands was now drifting away.

Here, the unifying thought is "loneliness" or "abandonment." The purpose of each sentence is to communicate the man's loneliness. But the thought is never stated. The paragraph makes the reader feel it, or deduce it, from the written facts.

Suppose you are reporting on some work that you consider highly significant. The unifying thought is "this work is of major importance." But to transmit that thought effectively to your reader, you must let him derive it for himself.

Carefully choose the facts that you feel will communicate the full meaning of the work. Keep firmly in mind the unifying thought of "significance" and start to write. That unstated unifying thought will connect the sentences of your paragraph.

I have shown you some ways of building your paragraphs around a single thought or action. There are many others. Once you have the paragraph's unifying thought firmly in your mind, you will find the organization that fits into your style, your subject, and the rest of the piece you're writing. Trust yourself. Your paragraphs will work, if you absolutely demand that every sentence contribute to the unifying thought.

Chapter 3
The Tools

How to Make your Writing Understandable

This chapter takes up the mechanics of writing—the tools we have to use to turn our thoughts into an understandable message on paper. It shows you how to use four essential tools of business writing: grammar, punctuation, spelling, and visual aids.

The first three columns, **Answers to questions,** contain my answers to readers' most frequent questions on grammar, usage, and organization.

A three-part series, **Punctuation: the voice behind the writing,** explains how punctuation marks translate the inflections of speech into writing and gives you rules for mastering commas, hyphens, and other cryptic marks of our language.

Two columns on spelling follow. **Yes: you *can* learn to spell** gives you eight basic rules for American English spelling. **How to remember the spelling of impossible words** provides four techniques for fixing the hard-to-spell words firmly in your mind.

Then we come to the appearance of your piece. **Writing helpful headings** demonstrates the role of headings as visual guideposts for the reader and explains how to write informative ones. The columns on illustrations—**Could you illustrate that?** and a five-part series called **A short primer on illustrations**—explain when and where to illustrate your text and how to use your word processing and presentation software to create pictures with maximum effect.

Answers to questions

Over the years, many of you have written to ask me if I can clear up some difficulties in English grammar, style, and usage. Since certain questions continue to recur, I thought everyone might appreciate a periodic column devoted to tackling these recalcitrant problems.

If you have a question on a particular aspect of English or business writing, please do send it to me. I'll try to answer it in one of the next "Answers to Questions" columns.

Question: When do I use "affect" and when do I use "effect?"

Answer: This is one that bothers most people most of the time. Here's the story.

There are *four* words involved here, each with a different meaning.

The **verb** "to affect" means "to influence," "to move emotionally," or "to pretend."

- "Adding titanium oxide **affected** the solution dramatically: it exploded."

The **verb** "to effect" means "to make" or "to accomplish."

- "This policy should help **effect** a transition from communism to free market."

Now, hold on to your seats. Here's the next part.

The **noun** "affect" is a psychological term used mainly by people in, or affecting to be in, the medical profession. It means, roughly, "the condition produced by an emotion."

- "After the shock, her **affect** was completely blank."

The **noun** "effect" means "result."

- "The **effect** of adding titanium oxide to the solution was a gigantic explosion."

The best way to sort your way through this maze is to figure that the **verb** you usually want is "to affect," the **noun**, "effect." You want either **to affect** something or to produce **an effect**.

Question: What about "who" and "whom?" Is "whom" just the word you use when you want to sound impressive?

Answer: No! "Whom" is the object of a verb or preposition. "Who" is always the subject.

- "The client to **whom** you gave the contract is the one **who** went bankrupt last year."
- "**Who** came to see you?"

Particularly in written English, you must make the distinction.

Question: "What *is* a split infinitive, and why can't I split it?"

Answer: An infinitive is "to + verb:" *to laugh, to be late.* You split the infinitive when you put a word between *to* and the verb: "to *easily* laugh," "to *intentionally* be late."

Now grammarians will tell you not to split the infinitive because the infinitive is a whole that should not be broken up. As a general rule, this is true. However, clarity of meaning is much more important than strict grammatical accuracy. Today, a split infinitive is considered permissible if it is necessary for clarity, understanding, or force of expression.

- "At this stage of the experiment, it's important to **really** watch out for breaks."

Question: I remember learning that you should never start a sentence with "because." Is that so? And if so, why?

Answer: It's not always so. It is perfectly grammatical to start with "because" when you are writing a complete sentence:

- "**Because** we added the titanium oxide, the solution exploded."

The trouble arises when people begin with "because" and then write only a sentence fragment:

- "The solution exploded. **Because** we added the titanium oxide."

The problem is not "because," the problem is the sentence fragment. A sentence fragment is simply a statement that doesn't make complete sense by itself. It's a clause or phrase that needs something else to be complete.

"Because we're going to the moon" is not a complete sentence. It doesn't make sense by itself. It needs another clause to make the thought complete.

So you can start a sentence with "because." But the "because" clause must be followed by an independent clause (a clause with a subject and verb) that completes the sentence.

Because I have explained this, I'll stop here.

See?

Write me your questions and I'll try to answer them.□

Answers to questions

Thank you all for your great responses to my last "Answers to Questions" column. I received so many questions that I had to select the most frequently cited problems to tackle here. However, I promise to take up every question you send me in upcoming columns.

"Hopefully"

Q: "Can we have a column on 'hopefully,' hopefully soon?"

A: Here is the latest word on this source of purists' wrath: *Webster's Ninth Collegiate Dictionary* (the most recent edition) gives two definitions for "hopefully": (1) in a hopeful manner (He took the high jump **hopefully**.) (2) it is hoped (Can we have a column on "hopefully," **hopefully** soon!)

Now, *Webster's* points out that the order of its definitions is based on historical usage, not preference. **Hopefully** was used to mean "in a hopeful manner" before it was used to mean "it is to be hoped."

The writers of *Webster's* stress that Definition 2 is by now acceptable. You can use "hopefully" to mean "it is hoped," as in: Hopefully, we'll meet our budget.

This is the usage that has drawn so much critical fire, many purists claiming that only Definition 1 is correct. The compilers of *Webster's* strongly insist that both usages are acceptable. They point out that, although it still arouses occasional objection, the second meaning "has been in use at least since 1932 and is well established as standard."

And that, for now anyway, is that.

"Myself"

Q: When is it correct to say "myself"?

A: In two situations: (1) When you want to emphasize the pronoun I (I did it **myself**.) (2) When you are both the subject of the sentence and the object of a verb or preposition (I hurt **myself**. I worked it out by **myself**.)

If you limit your use of the word to these two cases, you'll stay out of trouble with it. The problems start when you're not both subject and object.

The most common mistake is to use "myself" instead of "I" or "me" when there is a multiple subject or object.

For example, few people would think of writing: He sent the memo to **myself**.

However, it is not uncommon to find: He sent the memo to the director, the research manager, the sales representatives, and **myself**.

This, of course, is just as wrong as "He sent the memo to myself." In both sentences, the correct pronoun is a simple "me."

Similarly, this sentence is clearly wrong: **Myself** accompanied the president.

However, many people write: My manager and **myself** accompanied the president.

Again, that's just as wrong. Both sentences require "I."

So—allow the word "myself" only an emphatic or reflexive role, and it won't give you any more trouble.

"Since"

Q: When you use "since," how do you indicate whether you mean "because" or "after the time that"?

A: Correct use of the comma makes the distinction—if the sentence doesn't start with "since."

This sentence is ambiguous: **Since** you left the group, we've been much busier. It could mean either: (a) We've been much busier **because** you left the group, or (b) We've been much busier **in the period** following your departure.

To avoid this ambiguity, don't start the sentence with "since." Find another expression, such as "because" or "ever since," to make your meaning clear.

Now, if you put your "since" clause later in the sentence, you can be completely unambiguous. Just remember to put a **comma** before "since" when you mean "because," as in: We've all been working 18-hour days, since PM No. 1 broke down. This sentence can have only one meaning: We've all been working 18-hour days **because** PM No. 1 broke down.

Omit the comma, and you have another meaning entirely: We've all been working 18-hour days since PM No. 1 broke down.

To keep "since" in line, then, put a comma before it when you mean "because." And don't start a sentence with it if the sense could possibly be ambiguous.

"Preventive" or "preventative"/"maintenance" or "maintainance"

Q: Which ones are correct?

A: Both **preventive** and **preventative** have been in use since the seventeenth century. However, most authorities agree that **preventive** is preferable today.

On the other pair, **maintenance** is the only option.

Those are all the questions I can answer this time, but keep writing!□

Answers to questions

1. How do I get started?
2. How can I make my writing look more professional?
3. How can I get my writing to flow better?

These three questions lead us to work on a critical goal of business writing:

To communicate a message in such a way that busy people will take the time to read it. Let's consider each one.

How do I get started?

We all have trouble writing the first line—for a very good reason. We know it's the one line the reader is sure to read. If it captures his attention, there's a good chance he'll go on reading. If it doesn't, he'll probably stop right there.

The challenge of the first line actually contains the secret of writing it. Remember, this is the one line you can be sure your reader will read. Therefore you must use it to do one of two things:

- Grab the reader's attention.
- State your main message.

The attention-getting start

The advantage of the attention-getter is that it piques the reader's interest and puts him in the mood to read on. If you find a natural attention-getter, use it. For example, it came naturally to me to start this column with the questions I was asked. Since they are universal questions, they tend to get people's attention. However, beware of a forced attention-getter; it will look strained and will turn your readers off.

The main message start

If an attention-getter does not come naturally or seem appropriate for the piece you're writing, state your main message first. If it's important to the reader, it will motivate him to go on reading. Even if it's not all that important to him, you will at least have got your main message across—since the first line is the one people are sure to read.

How do you decide on and state your main message? Close your eyes and imagine that you have only one minute with your reader. In just 60 seconds, he will dash out of the room to catch a plane. What do you want to tell him in that crucial minute?

Write down your answer. Don't worry about style or grammar; you can fix them later. Just write down your main message as if you were saying it. That's your opening.

What next?

If you begin with an attention-getter, write your main message next.

Once your main message is down (with or without an initial attention-getter), imagine that your reader has paused at the door. Your main message was so interesting, he decided it was worth missing the plane to find out more.

Now, what does he ask you?

Look at your main message again. What questions would it lead your reader to ask? Imagine him asking you those questions—and answer them.

Your answers will form the draft of the rest of your piece.

Try this method of drafting. It will produce a memo, letter, or report that tells the reader the things he needs to know in the order in which he wants to know them. That is the essence of a readable piece of writing.

How can I make my writing look more professional?

The problem with this question is the work "look." Good business writing doesn't "look" anything at all, because it doesn't draw attention to itself. It transmits the writer's message as accurately and understandably as possible.

Professional writing is clear, not vague. It uses simple English words, not pompous "businessese." And it is honest, not hedging.

Make your writing *clear*, *simple*, and *honest*— and you will make it really professional.

How can I get my writing to flow better?

Think of the difference between a "staccato" and a "legato" sound in music. In the chopped "staccato" sound, the notes are clearly separated, while in the flowing "legato" sound, they are connected.

It's the same in writing. If you always write just one sentence after another, with no indication of their relationship, your writing will be choppy. To make it flow better, add some connections between your thoughts.

You can do this by using transition expressions to link sentences and paragraphs. Here are some examples of transition expressions.

To add: also · besides · furthermore · incidentally · similarly

To show sequence: first (second...third) · later · then · finally

To explain: in particular · for example · chiefly · let us say

To contrast: however · instead · still · by contrast · nevertheless

To summarize: accordingly · consequently · in brief · therefore

Good luck and keep sending me your questions.

Punctuation: The voice behind the writing

Part 1

It's not a matter of dots and dashes, you know. Punctuation is the staff of writing, the power we pack behind the words we choose to communicate our meaning. It's what makes a sentence stand up, bend over, lean back, or march straight out at you. Used correctly, it prevents unintentional ambiguities; sprinkled thoughtlessly about, it can actually distort the sense of the sentence, conveying something entirely different from what you wanted to say.

What *is* punctuation? It is literally the translation of the inflection of speech into writing: the pauses, the emphases, the raised or lowered pitch you use to invest words with your own intent. It separates ideas, clarifies the logic of your expression, and gives your writing dynamics.

Technical writing in particular cries out for correct punctuation. Its readers are frequently presented with a good deal of "heavy" information in a short space. Proper punctuation makes this rich diet digestible—even palatable.

How important is it? Just look what happens without it:

The elemental composition of the micrograph is illustrated by a series of scans graph 2 overall composition graph 3 structure of debris and graph 4 background structure

When punctuated, this spiderweb of words is actually a straightforward sentence:

The elemental composition of the micrograph is illustrated by a series of scans: Graph 2 (overall composition), Graph 3 (structure of debris) and Graph 4 (background structure).

All right. To write effectively, you have to punctuate correctly. If you don't, you will either irritate your reader by holding up his flow of understanding or mislead him by writing something that you didn't mean, as in the following example.

The writer of a technical report began:

The 12-inch cleaners, which are described in this report, did not function well under stress.

But he didn't mean it! Further reading revealed that, contrary to this statement, many of the 12-inch cleaners functioned perfectly under stress. Only certain ones failed—the ones he was about to describe. What he *meant* was:

The 12-inch cleaners which are described in this report did not function well under stress.

A couple of commas changed the whole meaning.

Of all the twelve major punctuation marks, that little, power-packed comma gets us into the most trouble. Because it lends itself to legitimate stylistic variations, it often leads writers to forget that it *means* something (a brief pause) and to use or omit it more or less according to mood!

A few guidelines and a general rule should help you conquer the comma. Let's start with the guidelines.

- Use a comma to separate clauses in long compound sentences.

The specimen is removed from the solution of liquid nitrogen, and the sample module is screwed onto the device.

Here, the comma helps the reader digest two detailed pieces of information. In a shorter, less detailed sentence, the comma would be unnecessary:

The specimen is removed and the module is attached to it.

You *may*, however, use it for emphasis:

We have met the enemy, and they are ours.

- Use commas to set off a subordinate clause when the main clause could stand alone, unchanged, without it.

The technician, who works regularly for Doyt Enterprises, quickly explained the problem.

Here, the main clause—*the technician quickly explained the problem*—can stand alone. The subordinate clause, set off by commas, adds information but does not define or determine the main clause. If the meaning of the main clause *depends* on that subordinate clause, use no commas:

The technician who works regularly for Doyt Enterprises is the man you want to see.

The main clause would make no sense without the subordinate clause. It cannot stand alone.

- Use commas to separate items in a series.

Today, most stylists prefer to omit the comma before "and":

The mill manager, research directors, group leaders and sales personnel all went to see the president.

- Use commas to separate two or more modifiers of a noun if the noun could stand alone, without either.

He is a brilliant, hardworking engineer.

You could reverse the sentence to read:

He is an engineer who is brilliant and hardworking.

But look at this one:

He is a brilliant chemical engineer.

Chemical changes the meaning of *engineer*; it merges with it to make a new concept. You could not write:

He is an engineer who is brilliant and chemical.

If you're ever in doubt, this reverse-the-sentence test will tell you when and when not to use a comma to separate modifiers.

These few guidelines show you that commas should be used to separate items that *need* to be separated in order for the reader to understand the message. Like all the other punctuation marks, the comma must always be used to mean. It must never be added simply to "break up" a sentence after a certain number of words or to give a visual impression of variety. This leads me to the general rule for commas:

Consider a comma where you would pause briefly in speech (not in thought). Then use it only if it seems necessary to clarify, and sufficient to separate, your written ideas.

Next time, we'll see how this rule can help you avoid all major comma-blunders and how you can make the other punctuation marks work for, not against, your meaning.

Punctuation: The voice behind the writing

Part 2

Punctuation reflects inflection. Remember that and most of your battles with the 12 "magic markers" will be half-won. Never again will you find yourself aimlessly sprinkling a letter with commas and semicolons—not unless you are trying to reflect the halting speech of one grappling with English for the first time.

In the last column, I gave a general rule for using commas to save you from the three most common comma blunders. The rule, again, is:

> Consider a comma where you would pause briefly in speech (not in thought). Then use it only if it seems necessary to clarify, and sufficient to separate, your written ideas.

Now let's look at the comma blunders you won't make any more.

1. Blunder: The illogical interpolation of commas, either to break up a long sentence or to reflect a pause in *thought*.

Example: *Those who believe the oil price will continue to hold steady, argue that all the major adjustments have already been made.*

Explanation: Here, the writer probably paused in his mind to think about the people who held that belief. Then, remembering his subject, he went on to complete the sentence (. . . argue that etc.). However, it is illogical in the extreme to unbind the two integral parts of the sentence, the subject and the verb. Instead of clarifying ideas by separating them, this comma obfuscates one idea by fragmenting it.

2. Blunder: The omission of one or more commas to set off a clause.

Example: *The Fortune 1000 companies which depend on fast, accurate communications, all use some sort of word processing machine.*

Explanation: Although this may at first sight look like a No. 1 Blunder (random comma between long subject and verb), it is actually carelessness of a different sort. The comma after *communications* serves to confuse totally the identity of the subject of the sentence. Is it all the Fortune 1000 companies or in particular those that depend on fast, accurate communications? Since the writer did bother to insert the one comma, we may assume he meant to add its complement:

> *The Fortune 1000 companies, which depend on fast, accurate communications, all use some sort of word processing machine.*

Now there is no doubt that the subject is all the 1000. The one comma meant nothing; the two surrounding the subordinate clause set it off and free the subject for clarification.

If the writer had spoken the sentence to himself, he would not have made this mistake, for he would have found himself naturally pausing after both *companies* and *communications*. And he would have translated these pauses correctly into commas.

3. Blunder: The use of a comma between independent clauses.

Example: *The project seemed about to be approved, it was only a matter of time.*

Explanation: Commas indicate brief pauses within complete sentences. A comma is not sufficient to make the big break between two independent, if connected, thoughts. Here, the writer has several alternatives. He can simply make the two pieces of information two separate sentences:

> *The project seemed about to be approved. It was only a matter of time.*

This choice is clear and correct, but it establishes no close connection between the two expressions. To do so, the writer could use a semicolon, a dash, or parentheses. Look how the results differ:

> *The project seemed about to be approved; it was only a matter of time.*
> *The project seemed about to be approved (it was only a matter of time).*
> *The project seemed about to be approved—it was only a matter of time.*

The semicolon establishes a basic connection between the two ideas without adding any particular emphasis.

The parentheses weaken the second time, implying that it is of only marginal importance to the sentence as a whole.

The dash sets off the second idea, calling attention to it and adding a touch of drama.

You see how powerful these little marks can be? In this case, you would just choose the one that matches your style and intent. Always remember, however, that the comma marks only the briefest of pauses. It is not sufficient to indicate the separation of two self-contained clauses.

Now that you've got the elusive comma under control, you'll find the other punctuation marks easy to master. Next time, we'll set up some "signposts" to guide you through the remaining few difficult intersections and bumpy byroads.

"The Language of Business" is a periodic column on effective business communications.

Punctuation—packing in personality

Part 3

If you've been putting the comma rules into practice, you may already be enjoying freedom from yesterday's punctuation perplexities. Now you can start using those elusive marks to put personality into your writing. Remember—they're the translation of your unique voice behind the common currency of words.

To explore all the possible uses and pitfalls of punctuation, we would need at least one column per mark. Instead, let us look at the most common opportunities and problems of punctuation in technical writing.

Enclosers—brackets, parentheses, quotation marks

Barring a few other arcane appearances, brackets [ ] generally enclose references, parentheses ( ), added information. For example:

The procedure followed (one developed for the F-19 project but adapted to fit present purposes) showed a great variance in molecular distribution [Fig. 23].

The distinction is clear. However, the rule I will suggest offers a little more trouble:

Brackets and parentheses intrude. Use them as little as possible.

Too often, we start writing but get suddenly waylaid by a tangential idea that rushes in to be included. What do we do? Stick it in parentheses! The parentheses then reflect uncontrolled, unfocused thought and *deflect* attention from the main message.

In the above example, the writer's main message is that the procedure showed a great variance in molecular distribution. But the sentence contains two other messages as well. One, the procedure was developed for another project and adapted for this one. Two, the reader should search for Figure 23. The two enclosers are too distracting.

When parentheses threaten to appear, see if you can't make the expression:

- a coherent part of the sentence, clearly related to its message,
- another complete sentence, or
- a footnote.

If none of these work, ask yourself why you need the expression at all.

If you have to use parentheses, keep the information in them short. And remember that other punctuation marks go *outside* the parentheses if they refer to the sentence as a whole, *inside* if they belong to the parenthetical expression.

Quotation marks offer an opportunity to inject a new voice into your writing, giving it more variety and verve. Feel the difference between the direct quote:

The president said, "Let's get some more men on our team!"

and the lackluster indirect form:

The president said that we needed more men on our team.

Quotation marks are used to indicate all quoted matter as well as unusual words. Punctuation with quotation marks defies logic. As with parentheses, all marks *but the period and comma* go inside if they refer to the quotation, outside if they belong to the surrounding sentence, if any. Quixotically, the period and comma *always* go inside the quotation marks, whether or not they refer to the quotation itself:

Did Einstein actually say it was "all relative"?

Einstein didn't actually say it was "all relative."

Connectors—semicolon, dash, colon, apostrophe, hyphen

The semicolon and the dash are alternatives to the period, used to indicate a close relationship between two independent clauses. The semicolon merely establishes that relationship. The dash is an attention-grabber, a rein-puller, an imperious gesture that points dramatically to the idea to follow. Remember that dash rhymes with panache, and you'll know when to use it.

The colon is simply an arrow. It points unemotionally to information that proves, details, or fulfills the statement made. Use it before lists, definitions, long quotations, and any time you are really translating the words, "the following," into a punctuation mark.

The apostrophe can tighten your discourse by its quick expression of possession. *The technicians' reports* reads better than the mouthful: *the reports of the technicians*. Just remember to place the apostrophe *outside* plural forms so as to distinguish them from the singular: *the technician's reports*. If you've got a compound expression, the apostrophe follows the last part only: *the managers and directors' lunch*.

And finally, the hyphen. If you're not sure whether to use it and your dictionary doesn't help, ask yourself one question: Would the expression be confusing or difficult to read without the hyphen? If so, use it. Here are some examples of confusion cleared up by the little hyphen:

chemical containing polymers
chemical-containing polymers

a light weight lifter
a light-weight lifter

infrared
infra-red

reentry
re-entry

Conclusion

Good writing creates coherent, logical patterns and conclusions out of a barrage of thoughts. Enclosers can interrupt this process, translating instead the randomness of thought into confusing messages. Connectors can do the opposite by establishing relationships between thoughts. Used knowledgeably, these and all other punctuation marks will lend not only clarity but voice to your own writing.

Effective business communications is the goal of "The Language of Business" column. The author welcomes your comments and suggestions for this feature.

Yes: You *can* learn how to spell

Q: "Help! My spelling is atrocious. Is there any way I can learn how to spell?"
A: "Yes!"

Lately, this is one of the most frequent questions I have been asked—usually in despair. Yet despite its apparent capriciousness, English does have a few guidelines for spelling. There are also techniques for remembering difficult words.

Here, we'll look at some guidelines for American English spelling. Next time, we'll take up techniques for remembering the spelling of the words that still return to haunt you.

1. When you add a suffix, retain the *sound* of the original word.

This rule applies particularly to **c**, **g**, and single vowels.
The consonants **c** and **g** have a hard sound (**cat**, **gum**) *except* when they appear before **e**, **i**, or **y**. Then, they have soft sounds: **c** sounds like s (place,Cyrus), and **g** like **j** (**huge**, **raging**).

Now, when you add a suffix, you must keep the hard or soft sound of the original word.

To keep the soft sound, **c** or **g** must be followed by an **e**, **i** or **y**: notice + able = noticeable; manage + able = manageable. If you dropped the **e**, you would get noticable (pronounced noti**k**able) and managable (pronounced with a hard g).

To keep the hard sound, **c** or **g** must be separated from a suffix beginning with **e**, **i**, or **y**. Usually, this means putting a **K** after the **c** (picnic + ed = picnicked) and **doubling** the **g** (sag + ing = sagging). If you didn't add these letters, you would get picnicing (pronounced picnising) and saging (pronounced **sayjing**).

Now we come to the vowels. These rules refer to words that have the **stress on the final syllable**. Again, the rule is that the sound of the original word must be retained.

If a single vowel is followed by a single consonant, the sound is short (**can**, **met**, **big**, **top**, **hug**, **occur**).

If a "silent e" follows that consonant, the sound is long (**sale**, **recede**, **like**, **spite**, **note**, **prove**, **huge**).

To keep the short sound before a suffix beginning with a vowel, **double the final consonant**: can + ed = **canned**; big + er = **bigger**; hug + able = **huggable**; occur + ence = **occurrence**.

To keep the long sound when adding a suffix beginning with a consonant, **keep the "silent e"**: huge + ness = hugeness; spite + ful = spiteful.

Before suffixes beginning with a vowel, such as "**-able**" or "**-age**," you **drop** the "silent e" in most cases: note + able = **notable**; prove + able = **provable**; like + able = **likable**; sale + able = **salable**. There are a few exceptions, such as mileage (mile + age) and lineage (line + age).

2. It's "i" before "e" except after "c."

This old rule works for words that have an "ee" sound: **shield**, **believe**, **mischief**, **conceit**, **ceiling**, **perceive**. Words with other pronunciations are spelled "ei": **neighbor**, **counterfeit**, **foreign**.

Exceptions: either, leisure, seize, sheik, weird.

3. For words ending with a consonant + "y," change the "y" to "i" before a suffix.

For example: beauty + ful = beautiful; lady + es = ladies

4. Most words ending with the sound "-cede" are spelled "-cede."

There are just three common "**-ceed**" words: **exceed**, **succeed**, **proceed**. The only common "**-sede**" word is **supersede**.

5. Most words ending with the sound "-ize" are spelled "-ize."

Only two common words end in "**-yze**": **analyze**, **paralyze**. Other "sets" to be considered by themselves are these:

- **"-mise" words:** surmise, demise, compromise.
- **"-prise" words:** surprise, comprise, etc.
- **"-rise" words:** sunrise, moonrise, uprise, arise.
- **"-vise" words:** advise, supervise, revise, improvise.
- **"-wise" words:** otherwise, likewise, etc.

6. Most words ending with the sound "-cy" are spelled "-cy."

The few exceptions include: courtesy, ecstasy, fantasy, heresy, hyprocrisy, to prophesy (the noun is prophecy).

7. Most words ending with the sound "-ence" are spelled "-ence."

Only four common words end in "**-ense**": **defense**, **license**, **offense**, **pretense**.

8. Most words ending with the sound "-ify" are spelled "-ify."

Only two common words end in "**-efy**": **liquefy** and **stupefy**.

I hope this has been clarifying, not stupefying. Don't try to memorize it all at once. Just keep this page as a help and reminder. Next time, we'll look at some ways to fix unruly spellings firmly in your mind. ☐

How to remember the spelling of impossible words: Four techniques

Are you a good speller? Before you go on reading, see if you can pick out the correctly spelled words from the following list. (The answers are at the end of the column—but don't deprive yourself of the game by going there first. You might even turn this into a serious office contest, after work hours.)

absence/abscense	harebrained/hairbrained
accessible/accessable	insistent/insistant
auxilliary/auxiliary	irresistible/irresistable
cancelled/canceled	licence/license
committee/commitee	liquefy/liquify
coolly/cooly	occasion/occassion
correspondence/correspondance	occured/occurred
criticise/criticize	permanant/permanent
dependant/dependent	recommend/reccomend
develop/develope	separate/seperate
dispensible/dispensable	sieze/seize
excede/exceed	sieve/seive
embarass/embarrass	simplefied/simplified
an envelop/an envelope	supercede/supersede
federel/federal	withhold/withold

If you haven't won any national spelling bees lately, you probably had at least a few doubts along the way. These words come from a list of the most commonly misspelled words in American English today. How do you fix these slippery spellings in your mind, once and for all?

To start, look back at the spellings you missed in the list above or note down words you know you often misspell. Now pick out no more than five. Don't try to work on more than five at a stretch, or you'll get muddled.

You can use the first method for every word—and it's often all you'll need. However, you may have to supplement it with one of the others. Try one or two of the methods on each of your five words today. By tomorrow, you'll have five fewer things to worry about in life.

1. Look up the word in the dictionary.

Spell it out to yourself. Then write it down correctly 10 times. Yes, 10 times. I know it sounds like "detention time" in school, but somehow, the act of writing it properly over and over again fixes the word in your brain. You fix it kinesthetically as well as visually.

Have you done it? Ten times?? Then close your eyes. Try to see the word spelled correctly. Spell it to yourself. Then look at the word. If you made a mistake, write it out correctly five more times.

2. Separate the word into syllables and pronounce each clearly to yourself.

This will work only for certain words, such as: com mit tee (pronounce: **kom mit** tee), cool ly (pronounce: **kool ly**), em bar rass (pronounce: em bar **rass**), rec om mend (pronounce: rec **om mend**), with hold (pronounce: with **hold**). For words in which the spelling makes very little difference to the sound—such as **absence** or **accessible**—try one of the methods below.

3. Use another form of the word to remind you of the correct spelling.

Say you have trouble with the word *absence*. Look it up in the dictionary and find another form that you know well, say **absent**. Since you have no problem remembering that **ab sent** is spelled **s-e-n**, imagine *absent* when you have to write *absence*. The adjective will help you remember how to spell the noun. Similarly, for **exceed** you could imagine exc<u>ee</u>dingly; for **federal**, feder<u>a</u>ted; for **simplified**, simpl<u>i</u>city.

4. Pull out the letters that give you problems and make up an association with them.

If you learned one somewhere, use it. Otherwise, make up one of your own. Whatever you use, be sure it's easy to remember—easier than the word itself! Here are some that have helped me:

accessible: if it's not h<u>i</u>dden, it's access<u>i</u>ble.
dispensable: The ph<u>a</u>rm<u>a</u>cy has dispens<u>a</u>ble drugs.
embarrassed: I get <u>r</u>eally <u>r</u>ed when I'm emba<u>rr</u>assed.
an envelope: I h<u>ope</u> there's a check in the envel<u>ope</u>.
license: To drive well, you need sen<u>s</u>e as well as a licen<u>s</u>e.
seize: You have to <u>se</u>e to <u>se</u>ize.

Spelling is important. Unfair as it may be, readers often judge the quality of your work by the quality of its presentation. If you're careless about your spelling, you can do yourself a serious disservice. Your word processor won't catch all possible errors (it accepts both **here** and **hear**, **there** and **their**, etc.), and you do occasionally write something by hand.

So take the time to fix those few problem words in your mind, now. It's worth it. □

ANSWERS: absence – accessible – auxiliary – canceled – committee – coolly – correspondence – criticize – dependent – develop – dispensable – exceed – embarrass – an envelope – federal – harebrained – insistent – irresistible – license – liquefy – occasion – occurred – permanent – recommend – separate – seize – sieve – simplified – supersede – withhold.

Writing helpful headings

Headings are technical writer's forgotten tools. In most reports, they have been whittled down to a standard repertory of five:

- Introduction
- Summary & Conclusions
- Recommendations
- Discussion
- Background Information.

Now, this makes a good organization for a technical memo or report. In the "Introduction," you give the busy reader an overview of what your report is all about and why he should read it. The "Summary & Conclusions" show him where the information contained in the report led you, and the "Recommendations" give him a clear idea of what you want from him. All this should cover some two pages, at the most. The remaining two sections flesh out that skeleton.

Fine. Keep that format—but only in your mind. Remember that headings can be powerful. They stand out. Thoughtfully worded, informative headings can:

- Give your reader the essence of your report by *highlighting* the important facts. This will please harried executives and make them look forward to your future missives.
- Give the reader enough intriguing information to make him eager to read the section carefully.
- Pull out your salient points for the reader, thereby directing his attention where *you* want it.
- Help you organize your thoughts.

Headings can be highlights

Instead of falling back on the standard, boring repertory, you might take a section heading from the most important sentence of that section. For example, suppose you are writing a technical memo on a new refining treatment. Write your introductory section, and then look for the crucial "topic sentence" or central concept. The new heading you use to replace "Introduction" might be:

AB Labs test a new XYZ refining treatment designed to increase mill production 45%

In place of a really useless word, "Introduction," you have now produced a heading that informs and interests the reader. From just that head, he knows *what* you tested (the XYZ refining treatment), *why* (to increase mill production), and precisely what *results* you hoped for (a 45% increase in production).

The next section, once titled "Summary & Conclusions," might instead be headed:

Tests indicate that, with certain modifications, the XYZ treatment could boost mill production 50%

Now, on a single page, you have told your reader what you were testing, why, and the preliminary results of the tests. You have also indicated that you will go on to explain the required modifications in the text to follow. *And* you've given your reader a bonus that he is sure to love: actual figures. All this in two short headings—which you might have lamely labeled "Introduction" and "Summary & Conclusions."

Good headings help you, too

Certain heads need to be more general than others, of course, depending on the length and breadth of the section they cover. But in writing the headings, you'll find you have a tool for testing the logical development of your text. Usually, the clarity of the headline reflects that of the text.

Take the section you used to call "Background Information." If you can't make its head any more specific than that, you've probably got too much unrelated background information all jumbled together. For the report on refining treatments, you may have written: a paragraph on similar studies made in your mill . . . another paragraph on the results of the XYZ treatment in other mills . . . another one on the recent history of refining at this particular mill . . . and yet others giving more details of the present study. As a matter of fact, this "outline" is typical of the background information sections in the technical memos I have seen.

Such a hodgepodge does not make for easy, interesting reading. On the contrary. So—break up your unenticing mass of background information into smaller, digestible sections. You might head them simply:

- Further details on the XYZ treatment
- Other refining methods used at AB mills
- Related studies on refining conducted here.

You *can* keep the umbrella head, "Background Information," and use these as subheads for subsections. My question is: Why? You would have to add a paragraph to introduce the subsections to follow, thereby creating additional text for the busy reader—who doesn't need it.

Writing helpful headings requires more effort than plugging in the old "Introduction–Conclusions" set. But you'll find the new heads help you as well as your reader. They sort out and classify your ideas. They keep your writing logical and focused. They make your writing more readable.

It's worth the effort. □

Could you illustrate that?

Imagine having to explain the difference between a rose and a tulip to somebody who has never seen either. You could delve into a detailed description of petal shapes, sizes, and geometrical arrangements — or you could achieve your goal much more quickly and effectively using drawings or photographs. The same is true for a piece of machinery or a production process with many unit operations.

Good illustrations can add both clarity and interest. Consider using an illustration when:

- Something seems difficult to describe briefly in clear, simple English.
- You're talking about numerical data (more than three or four items) that are related in some systematic way.
- You think your readers will have difficulty visualizing what you're talking about.
- Your discussion needs a strong reference point, such as a drawing with clearly identified parts, or a summary list of your most important points.

Yes I could — but I won't

An illustration highlights your point. The obvious question is: Does that point you're about to illustrate NEED to be highlighted? Clearly, you can't afford to highlight a minor point at the cost of your main message. Here are some simple rules for when to stay away from an illustration:

- Don't insult your readers by illustrating the obvious.
- Think three (3) times before trying to be funny or clever (that famous cartoon!).
- Don't turn a simple, short list that should be part of the text into a separate figure, unless you need to refer back to it repeatedly.
- Don't highlight a minor point with a figure.

As always, the cardinal rule is: *Use an illustration only if it will help get your message across.*

Picking the right tool

Your main tools are tables and figures. Tables collect nonpictorial data in columns, under appropriate headings. Everything else is a figure. Figures may be charts (pie, bar, curves, flowcharts), photographs, or drawings. Next time you open *Business Week* or a popular science magazine, take a look at the illustrations. You'll pick up many ideas on how to produce clear, helpful figures.

With all those tools available, which should you choose? You guessed it: whatever type will be most helpful in getting your message across. If you want the reader to understand some troublesome trends that call for action, you'll use a trend curve, bar chart, or similar figure. To help your readers visualize equipment or a process, a drawing or photograph would be appropriate.

To plan is to win

The best illustrations will be wasted — or worse, will hurt your message — unless you make them an integral part of your communication. You can do that only if you *plan* for it. Many writers include illustrations as an afterthought. As a result, they forget to refer to them in the text, leaving the reader to wonder about the purpose of all those elaborate diagrams.

So plan from the start. Study your outline. Decide where you will need illustrations, and what kind. Then make preparations to *use* the illustrations to their full advantage. If you decide to use a flowchart, you may want to organize the discussion to parallel the "flow" depicted in the chart. Make sure your outline reflects that decision.

Making it truly helpful

You'll have to attend to three elements: (1) the title or caption, (2) the body of the figure or table, (3) labels and explanations.

Title. Use clear, short titles that describe *what* you're showing and *why* you're showing it. Consider using full sentences: "Figure 3. Treating the sheet with XYZ compound results in a 7% increase in wet strength" vs. "Exhibit A. Wet strength of treated and untreated stock." If there are several illustrations, *number* them to allow clear, short text references.

Body of the illustration. Make your illustration neat and easy to follow. Have you really picked the appropriate type of illustration? Or would a grouped bar chart be clearer than that spider's web of dashed and dotted curves? Your main rule is to KEEP IT SIMPLE. Don't get trapped by details that are of no interest to your readers.

Labels, scales, and explanations. Use clear column headings for your tables and simple, helpful labels on your curves or diagrams. Make sure all units are clearly defined and all abbreviations spelled out somewhere. In some cases, you can improve a figure by adding explanatory text — say, a brief account of the major steps in a production process. The ideal illustration explains itself, so the reader can study it quickly even before reading your text, and still get a good idea of your main message.

Finally, make sure you refer to each figure and table, by its *correct* number (even after your last revisions), and place each illustration close to (but not before) its text reference.

Can you think of a better way to put your readers into the picture. □

A short primer on illustrations

Part 1: A general overview

You've generated lots of data, and your analysis has tweaked stunning trends and meanings from that silent mass of information. Now how do you convey your insights to your readers? Usually, you need some good illustrations.

Fortunately, it's very easy today to produce professional-looking illustrations, thanks to affordable graphics programs. However, "professional-looking" is not synonymous with "good." A good figure is one that grabs your reader's attention and conveys the essential information in a few seconds, demanding almost no effort. By that definition, many professional-looking visuals are bad visuals. And even if you have truly fine illustrations, they're useless if you put them where most readers will not look at them.

This time, I'll give a general view of what it means to illustrate your reports and memos effectively. In the next columns, we will go into details on how to do it, using Harvard Graphics and WordPerfect as a representative software combination.

Your main problem: distractible readers

Let me tell you some shocking things about your readers:

- **Fact 1: Even when they're interested, they are impatient and easily distracted.** They will skip sections and whole pages, and they'll give up altogether at the slightest provocation.
- **Fact 2: They'll never look at Figure 4 in the Appendix when you tell them to look.** If they study the figures at all, they'll go through the whole lot in one sitting.
- **Fact 3: When looking at a figure in the Appendix, they won't remember what the text said about that figure.**

What does that mean for you? First, you must **put your important illustrations into the body of the report,** not into the Appendix. Second, if you do leave some figures in the Appendix, they must be **self-contained** so the reader doesn't have to jump back to the text to understand them.

I know, it seems like a really tall order. "Illustrations right in the text!" you say. "Give me a break! I'm not a desktop publisher!" Actually, it's not at all difficult to do, as we'll see next month. (For instance, WordPerfect will import any figure you've created in Harvard Graphics or some other common graphics package and place it anywhere in your text, in whatever size you want.)

Think about it. Can you imagine reading a book in which all figures are lumped together at the end? You would write the publisher off as crazy. Yet we expect all our super-busy readers to digest foot-high piles of reports that are put together in just that way.

Of course, there may be some in-house complications. The communications group may not like the idea of integrating figures into the text, because it slows down the "report production line." If production is handled by secretaries, they may not feel like grappling with such advanced word processing. In either case, offer to show how easily it's done. You'll convert them when they see how simple it is and how much better it looks.

Creating figures that grab, hold—and release quickly

Presentation software truly has come a long way. If you often need to draw figures (curves, pie charts, bar charts, or freeform drawings) and are not using up-to-date software, take the plunge. (You can get Harvard Graphics 2.3, for instance, for less than US$ 300 from discounters.)

The biggest trouble with presentation software is the "default" options. Everybody seems to rely exclusively on these defaults—even though they are far from ideal. In most cases, you should take a few extra minutes to **customize** your figures: choose a heavy line style for your key variable; label lines and bars directly, not through a legend; delete unnecessary grid lines; and add explanatory notes to make it a "whole story."

Finally, don't be afraid to use **drawings** where they would be helpful. You may think of yourself as the "ultimate nonartist"—but today's software offers so much help that even I can produce acceptable drawings quite quickly.

Next time, we'll start looking at how to create compelling illustrations in Harvard Graphics and then put them right into a WordPerfect document.

A short primer on illustrations

Part 2: Integrating illustrations into your text

Last month, I discussed some general requirements for effective illustrations. The main obstacle you need to overcome is the extreme distractibility of your busy readers. That leads to three immediate guidelines: 1) place your key figures right in the text so people can study them as they follow your discussion; 2) don't just use all the default options of your graphics software but customize your figures so they convey your points as strongly as possible; and 3) add simple drawings and symbols where they would enhance your message.

Let's now see how to integrate your illustrations right into your text, using WordPerfect 5.1 and Harvard Graphics 2.3. These two are the most popular programs, but even if you use other packages, you may be able to follow the instructions with minor changes.

Generally, putting your figures into a WordPerfect document involves saving (or "exporting") each figure in an intermediate format that WordPerfect can read, then "importing" it into WordPerfect. WordPerfect imports figures by converting them to a special WordPerfect Graphics (WPG) format. The problem with this process is the *size* of the converted files (about 250,000 bytes per figure). Adding 10 figures to a WordPerfect document takes up 2.5 megabytes on disk (so you couldn't save it on a floppy disk, for instance). However, WordPerfect's "Graphic on Disk" feature allows you to save each imported figure separately from the main document.

What to do in Harvard Graphics

Harvard Graphics 2.3 offers you two major "intermediate" formats for exporting: Encapsulated PostScript (EPS) and Hewlett-Packard Graphics Language (HPGL). WordPerfect accepts either format and converts it to its own WPG format. I have found that the final print quality is better with the EPS format, so I suggest you use it. Here are the steps to follow:

1. Draw or retrieve your figure, save your final changes, then return to the Main Menu.
2. At the Main Menu, choose "Import/Export."
3. At the "Import/Export" menu, choose "Export Picture," then set the proper Directory, Picture Name, and Format options.
4. Under "Directory," choose your WordPerfect directory.
5. Under "Picture Name," type a name for the converted file (it should have the extension ".EPS").

1. Harvard Graphics figure imported into WordPerfect at reduced size

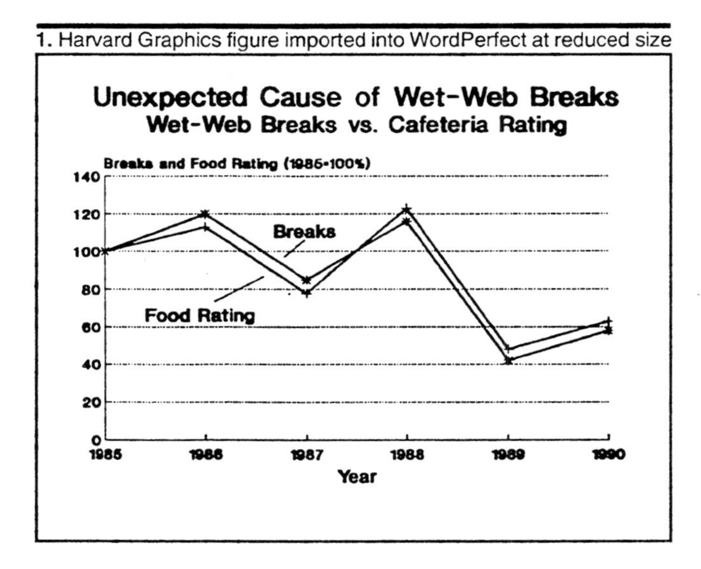

6. Under "Format," choose "Encapsulated PostScript," even if you are printing to a Hewlett-Packard LaserJet.

What to do in WordPerfect

You can find complete instructions in your WordPerfect manual under "Graphics, Create" and "Graphics, Define a Box." Here are the main steps:

1. As you finish typing a paragraph that refers to a figure, press ALT-F9 ("Graphics"), then choose 1 ("Figure") and again 1 ("Create").
2. At the "Definition: Figure" menu, enter the filename of your converted EPS figure from Harvard Graphics.
3. Define "Contents" as "Graphic on Disk" (=2) to keep the graphics disk file separate from the text file.
4. Enter a caption if you want to (numbering is automatic, but you can override that).
5. Set the horizontal position (left, right, or center).
6. Set either horizontal or vertical size (WordPerfect will adjust the second dimension automatically).

To delete a figure, you have to press "Reveal Codes" and delete the "Fig. Box" code. You can change many options (e.g., border style, outside and inside border space, caption style, and position of caption) for any figure by pressing ALT-F9 ("Graphics"), then 1 ("Figure"), then 4 ("Options").

Figure 1 shows a Harvard Graphics chart I imported into WordPerfect in this way. Isn't it nice not to have to turn to "Appendix 1: Figures" to find it? ☐

A short primer on illustrations

Part 3: Creating strong charts with Harvard Graphics®

Last month, you saw how to integrate key figures into your word processing document, to enable your readers to look at them easily as they read the text. I assumed that you were using Harvard Graphics 2.3 for figures and WordPerfect 5.1 for text, but much of what we are discussing in this series applies to other good software packages.

Now let's see how to create Harvard Graphics charts that get your point across as strongly and quickly as possible.

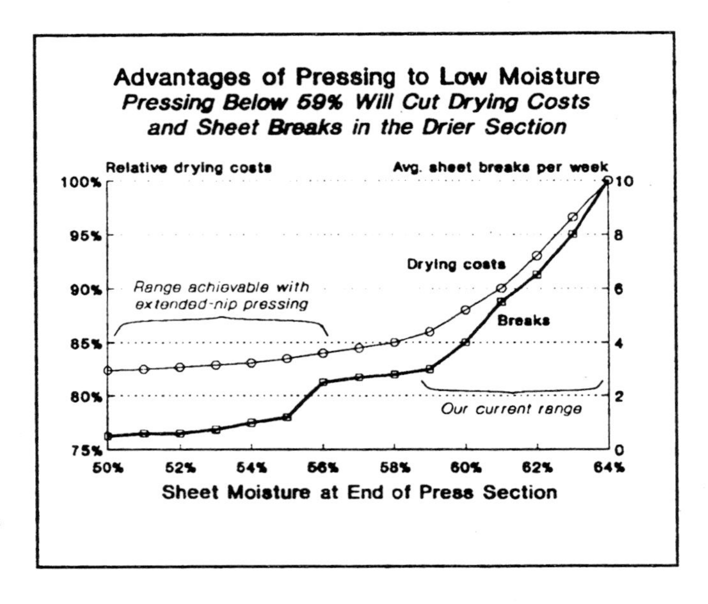

Three principles of powerful illustrations

Good illustrators use three simple principles in every figure:

- Make it self-contained: let it *tell the story at a glance.*
- Make it instantly absorbable.
- Make it attractive. Avoid clutter and other distractions.

Applying these principles means customizing your charts, rather than accepting all the convenient default options of your graphics program. I have some specific suggestions.

Label curves or lines directly rather than through a legend, unless this would make the figure too crowded. By default, Harvard places a legend below the x axis. To delete it, go to Page 2 of the "Bar/Line Chart Titles & Options" form and select "None" under "Legend Location." (If you decide to keep the legend, you can change options here to place it inside the chart, top or bottom, left or right.

To label curves directly, use "Draw/Annotate" (F4). First choose "Add," then "Text/" and press F8 to change the text size and other attributes. (The default size is 5.5; for curve labels, you'll want a smaller size—say, 3 or 3.5.) Then type the text and place it where you want it.

Label bars with values unless you have too many or want to show only rough trends. To see the trends, people can then go by the graphic image. If they want to know some specific value, they can find it immediately. To put values on bars, change the "Value Labels" option on Page 2 of the "Titles & Options" form.

Omit unnecessary grid lines and frames. For instance, if you label bars with values, you don't need the Y grid lines, and in most cases, a "half frame" (left and bottom line) is sufficient. You change these options on Pages 2 and 3 of the "Titles & Options" form.

Choose contrasting styles for lines and points. The default is line style 1 (a thin line) for all data lines. On Page 4 of the "Titles & Options" form, choose line style 2 (thick) for your key series. (Unfortunately, importing the figure into WordPerfect will somewhat decrease the thickness.) Consider dotted and dashed lines (styles 3 and 4) for projections or subordinated information. For data points, Harvard uses a dot, a cross, and a star for the first three series. You may find a square (style 4), diamond (6), triangle (7), or circle (9) more distinctive. (Appendix D in the manual lists all 13 options.)

Add explanations directly to the figure if you can find a place for them. Even if you put a figure directly into the text, explanatory notes help. They allow the reader to get the whole message in one place, rather than having to piece it together from various text passages and graphic elements of the figure. (For figures in an appendix, this suggestion becomes even more important.) To add textual explanations—plus lines, arrows, boxes, and braces—use "Draw/Annotate" (F4) or "DrawPartner" (CTRL-D).

Finally, be as informative as possible in your figure title and subtitle. Most people baffle their readers with cryptic phrases; yet there is no law against using full sentences in titles. For instance, the sample figure in this column uses the title and subtitle to tell the main point of the story. Two explanations in the figure spell out details. The curves are labeled directly; the line depicting the biggest benefit is thicker, and data points are represented by distinctive squares and circles. As a result, you can probably *get the story* quite quickly even without any accompanying text.

Next time, we will tackle the last task: turning you into the artist you didn't know you were. □

A short primer on illustrations

Part 4: Drawing for the nonartist

In the past, you may have hesitated to draw your own illustrations, because you were afraid they would look too amateurish. However, there really is no reason for such fears anymore. With today's graphics programs, your circles will be round, your squares square, your curves curved, and your corners neat and sharp. Adding shading is also a snap. Finally, there are many ready-made symbols to make the job even easier.

Let's see how to add pictures to your charts with Harvard Graphics 2.3. In general, use Harvard's Draw/Annotate (press F4) for simpler drawings. If you need to rotate or flip objects or do other sophisticated things, turn to Harvard's full-featured drawing program, Draw Partner (press CTRL-D).

1. Drawing combining and modifying Harvard Graphics symbols

2. Turtle drawn in Harvard Graphics, using a semicircle as a guide

Working with Harvard Graphics symbols

The easiest way to incorporate drawings into your charts is to use ready-made symbols from the Harvard Graphics symbol files. From the Draw/Annotate main menu, choose "Symbol," then "Get," then the name of the file that contains the symbol you want. Once you've retrieved the symbol into your chart, you can resize it, move it, change colors or patterns, delete or change portions of it, and add lines or shapes to it.

Figure 1 shows an example combining a light bulb symbol with a "businessman" symbol. I added eyes, mouth, and "idea beams." Then I deleted the face of the businessman, reduced the rest, and moved it into place. However, I had trouble getting the collar to fit. So, I kept only the tie and drew the collar and the shoulders with the "polyline" tool, using the symbol remains as a rough guideline.

This illustrates the main difficulty you'll encounter when working with symbols. It's easy to add to them, but not so easy to change their shape. If you need to change shapes, try using the symbol as a guideline, then delete it, leaving only your own drawing in place.

Producing original art

The problem with ready-made symbols is that they rarely seem to fit your needs exactly. For instance, when I wanted to illustrate the concept of "being in a tight spot," I thought of a turtle on its back. What could be easier than taking a "turtle symbol" and flipping it on its back? Of course, there was no turtle in the symbol library, and it really wouldn't do to have a pig or an elephant! So, I had to draw my turtle from scratch.

Figure 2 shows the fruits of my 15-minute labor. There are two points worth noting about it.

First, I did not use a mouse but only the cursor keys. This may be a little slower, but it works perfectly well. (Good news for those of you who hate mice!) To make finer movements, just press the minus key; each time you do this, the cursor moves in finer increments. (For even finer control, choose "Zoom" from the "View" menu.) To return to longer cursor leaps, press "+" as often as needed.

Second, I used a semicircle as a guide, then drew over that basic shape with the "polyline" tool. When I finished drawing, I deleted the semicircle. Drawing around basic shapes this way will help you get good results more quickly. Of course, if a symbol is available as a guideline, that's even better.

How to find good visual ideas

Perhaps the hardest part of drawing is coming up with interesting visual ideas. Look at the concept you're trying to illustrate, then consider many different areas of life, such as nature, sports, hobbies, cooking, war, animals, work, or shopping. Is there anything that does, experiences, or exemplifies your concept in concrete ways? If so, think of a simple way to turn that into a picture.

You can learn a lot from advertising. Just go through a few magazines and notice how the better advertisers express concepts visually. Why not cut out some examples and start building a "Symbol Ideas" file? It will be a great tool to have the next time you're struggling to get a smashing visual idea!

Chapter 4
The Types

How to Write Excellent Memos, Letters, and Reports

Now, with all the tools and techniques in place, you are ready to apply them to specific pieces of business writing.

Notes for a memorandum shows you how to write succinct, readable memos. **How to write a nicely spoken letter** gives you general guidelines for effective business letters.

Writing a complaint letter is a three-part series that spells out a technique for getting the results you want when you have a problem. And **Writing under fire** is a two-part series showing you how to approach readers who you think will disagree with you or resist your ideas.

The trip report and **The technical report** are both three-part series that show you how to prepare, draft, and complete an interesting, *useful* report.

Notes for a memorandum

Memoranda are slippery creatures. They're not letters. They're not reports (with the exception of "technical memoranda"—fairly informal short reports which we'll discuss at a later date). They are, however, a big part of all business's daily life. In fact, they could be said to be an intimate record of it.

I remember a particular scene that repeated itself quite frequently during my days as a public relations account executive. An idea, certain to revolutionize our handling of an account and triple our billings, would leap into my head. With equal alacrity, I would rocket out of my chair and head down the corridor to enlighten my immediate superior. He, a kindly soul, would listen, nod encouragingly, and suggest that I take it to the head of our unit—the client services manager. I would be gratified—but my joys of discovery would abate. For I knew what the client services manager would say. He said it every time, albeit with the utmost cordiality:

> "Sounds interesting. Why don't you put it in a memorandum and send it along to me?"

And I would trudge back to my office, trying to think of something pressing I had to do—something that would indefinitely forestall that miserable moment of "putting it in a memorandum."

Memoranda! For years, until I came upon a way to outwit them, I despised them. I hated them because I never felt *I* was writing them! They were an official communication from A to B or a group of Cs—but the people involved were mysteriously missing. There was no comfortable "Dear ..."to begin or cordial "Sincerely ..." to end. You didn't even sign the things, except with an occasional cryptic monogram.

The rigorous impersonal heading:

TO: DATE:
FROM:
SUBJECT:

set the tone of Serious Official Business. My memoranda would go into Files, for posterity's study and headshaking. In fact, I have even written a Memorandum to Files alone.

And so, I would at last sit down to demolish my idea in a memorandum. I don't think I ever actually wrote: "Pursuant to the discussion *re* the possibilities inherent in..."—but I came awfully close.

My reaction was absurd but, judging from the memoranda I've seen, not unique. Of course, an official document need be neither officious nor impersonal. A memorandum should be just what its name implies: a reminder, a source of reasonably complete and pertinent facts relating to a subject that may affect or interest the reader. When my client services manager asked me to "put it in a memorandum," all he wanted me to do was to put my thoughts down on paper—coherently, if possible—so that first I and then he could see if they were really worth his time and attention!

Maybe if we spoke English and called the things "reminders" instead of their more austere Latin appellative, they would cease to inspire awe in the writer and dread in the reader. Until that unlikely day arrives, though, we can make our own memos an easy, effective business tool. We begin with two clarifications.

First, why do you write a memo? Generally, you want to get some *specific information* to a *specific person or group* for a *specific purpose*.

Second, why *write* a memo? Putting your information in writing can *save you time—especially if you have a number of people to contact — organize* your thoughts and data, and provide a *record* of your facts or requests for others to refer to as needed.

With those two points clarified, we can address the final question: how do you write an effective memo? The system I evolved to circumnavigate my fear of memoranda continued to work for me long after I had put those particular dreads to rest.

I begin by making numbered notes, or reminders, for myself on a piece of scrap paper. The notes are all written out in full sentences. This ensures that I have a full grasp of the information or thought and am not hiding sketchy ideas behind sketchy shorthand. The method of making these numbered notes follows.

1. Write down all the facts or ideas you possess on the subject. Spend some time on this; it will free you of worries later on.
2. Write down the name of the person or people who will receive your memo. Include those likely to be copied on it.
3. Write down, in one or two sentences, how your information or idea can or does affect these people.
4. Write down, in one or two sentences, your purpose in sending them this memo, *i.e.*, how you want them to respond.
5. Write down, in one or two sentences, why you feel it would be to their advantage to fulfill your purpose.

Now, pull out Note Number 4. It forms the first sentence or paragraph of your memo. Number 3 forms the second sentence or paragraph. Number 5 forms the third. Take the pertinent information left in Number 1—that is, information that will be of use of interest to the people on your list—and transfer it from your sheet of scrap paper to your memo draft. List your data in order of importance to the people concerned (not necessarily to you).

You will find you have written a complete, well-organized, but agreeably unpretentious memorandum! You've stated your purpose, explained it, illustrated its importance to the reader, and given all the pertinent information needed. By the way, never assume there's no purpose in sending a particular memo. No one puts a list of facts together and sends them out for the fun of it. Your purpose is what makes this memo a personal communication between you and the recipient. It is the heart of the memo.

This seemingly upside-down method works for me for a number of reasons. The personal note-taking frees me of thinking about how to make my thoughts official and allows me to get them down in their entirety and spontaneity. It forces me to take a cold, hard look at the value of my thoughts or information. It enables me to discover and isolate the parts relevant to my recipients. Finally, since it starts out as a reminder from me to myself, it evolves into a personal, coherent message from one person to another—with no impersonal pomposity to cloud the issue.

*Caution:*you must do the steps in the order suggested precisely because they do not mirror the final memo. If they did, you would soon find yourself skipping the note-taking and plunging straight into a conventional, unmanageable memorandum! By keeping the notes to yourself in a different order from the arrangement of the final memo, you will retain their fresh, untrammeled character.

Of course, you won't always be able to transcribe your notes word-for-word onto the memo. You may need transitions and occasional rewording. But don't get caught in too many instances of "thus," "therefore," "consequently it behooves us to notice," "in this connection," and so on. Your memo is now organized to connect naturally. Let it! And good luck with your future *"reminders."*

How to write a nicely spoken letter

This morning I received a letter from a company that, I learned, would from now on be responsible for insuring my good health—or lack of it. It began by referring obliquely to a "merger which had been effected" between my original health insurance company and the one behind this letter. Sentence #2 read:

This merger was proposed in direct response to M.H.P.'s loss of federal qualification as a Health Maintenance Organization (H.M.O.) and the rehabilitation proceeding involving M.H.P. brought by the Superintendent of Insurance of the State of New York following a finding by the Superintendent that M.H.P. was insolvent and that without outside management and financial assistance, continued operations could adversely affect its policy holders.

Now. Had I not been alarmed at the possible loss of a hard-won insurance policy, I would probably have given up right there. As it was, terrifying phrases of foreboding stood out from that unreadable sentence, with a haunting intensity: "loss of federal qualification," "rehabilitation proceeding," "insolvent," "outside financial assistance," "adversely affect policy holders." By the time I got to the happy truth—three paragraphs later—that this merger made me "a subscriber of H.I.P., the largest New York State certified H.M.O.," my joy was muted, to say the least. I felt ill at ease and confused as to my status and that of whatever new monogram would now endorse my checks. And—the nagging doubt: if these people can't even put a legible sentence together in a letter designed to woo and calm a new (unwitting) subscriber—how clear are they about my rights and their responsibilities? Will I understand their claim forms? Should I still subscribe??

I mention this letter and my reaction to it to illustrate the enormous power and devastating effect a letter can have. It can lose you goodwill and business, simply by being jumbled and badly put together. The writer of this letter could have assured the company one continued subscription just by *organizing* the letter better—i.e., giving the good news first—by writing in *short, comprehensible sentences,* and by adopting a more *personal, friendly tone.* Letter-writing is, in fact, surprisingly easy, as you'll discover when you follow these few guidelines:

1. *Do not* start by thinking, "I'm about to write (dictate) a business letter." If you do, you will find yourself producing trite business cliches that will send your letter into the pile of uninteresting, unread mail written and received every day. Some of the most common "tired starts" to be avoided follow, with suggestions for more human alternatives:

- "Pursuant to your request, we enclose herewith the copy of . . ." Say instead: "I am happy to enclose your copy of . . ."
- "Further to our telephone conversation this morning, I take pleasure in forwarding . . ." Instead, "Here is the . . . that I promosed you on the phone this morning."
- "As per your request, enclosed please find . . ." Instead, "The . . . you asked for is enclosed."
- "We anticipate we will be in a position to fulfill your order . . ." Better: "We expect to fill your order by . . ."
- "We are forwarding under separate cover . . ." Better: "We have sent you the . . . by parcel post."
- "In reply to yours of March 12, please be advised that we are not in a position to . . ." Try: "Unfortunately, we cannot . . ."
- "We wish to acknowledge receipt of . . ." Instead, "Thank you for your letter."
- "Reference is made to your letter of May 4, which regrettably I have just had the chance to receive and answer." Why not: "Do please forgive me for being so late in replying to your letter of . . ."

Think of "a nicely spoken letter." Let your reader visualize you talking to him/her—and you'll write the right things!

2. *Do not* affect an excessive or insincere tone. For example:

- Unnecessary apology—"I'm sorry to take your valuable time, but . . ." If you were truly sorry, you'd come straight to the point without taking more of the reader's time to regret taking it. Or, you might not write the letter at all . . . .
- Excessive praise—"Your incomparable expertise in these areas . . ." Do you trust flattery? Would you be predisposed to believe in the sincerity of the writer who began with that phrase?

3. *Do not* repeat in detail the letter you are answering. Your letter will probably remind the reader why he/she wrote to you. You can mention the request by saying how you are responding to it. For instance:

Here are the tables you requested in your letter of July 19. I hope they will be useful to you.

All the information necessary for the reader—and a friendly, conversational note as well.

Now that you know what not to do, you can plan your letter.

1. *Do* plan to write a letter that is informative, short, and friendly. A long, wordy missive will be skipped through and possibly ignored altogether. But, remember, you are a person writing to another person. Your letter can establish a relationship of friendliness and goodwill—immeasurable values that a curt note may destroy. So, instead of writing something like:

Dear—, Our tests show that the addition of xeno causes delay in the hardening of the phenol mixture. Sincerely, —

add a few words and some humanity:

Dear—, We are pleased to let you know the results of the tests you requested. The addition of xeno did cause a delay in the hardening of the phenol mixture. If we can be of more help to you, just let us know. Sincerely, —

2. Ask yourself if the reader needs all the information you're about to give. If so, give it as simply as possible. If not, resist the impulse to show off your knowledge!

3. Be consistent in your salutation and signature. If you write, "Dear Joe," sign it, "Fred"—*not* F.H. Riordan."

4. Close with a pleasant remark, if possible relating to the subject of your letter. Instead of the worn-out "Thank you for your attention" or "Kindly advise," try "Thank you for taking the time to check this figure" or "Please let us know how you would like this sent."

The "secret" of good letter-writing is really very simple, then. You begin graciously, with a sincere statement of goodwill. You then go straight to the point, making your request or giving the information sought. Whenever possible, stress the positive. And close with a specific remark, either thanking the reader for something or giving a gentle reminder.

Then your letters will be read with attention—and goodwill.

Writing a complaint letter

Part 1: "The mad-rag"

Complaint letters are usually a disaster. The writer produces them in an incoherent rage. The reader gets insulted and misses the point. We would all much prefer never to have to write the wretched things. But—things happen to enrage us. To embarrass us. To make us simply have to complain.

The question is, of course, how we do it.

And that's usually the last question we ask ourselves when we're mad! We feel:

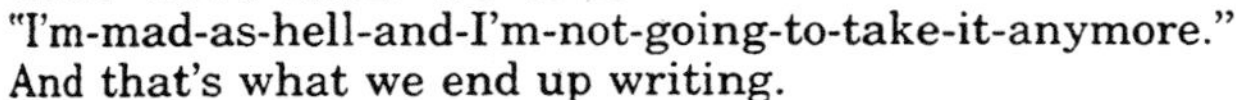

"I'm-mad-as-hell-and-I'm-not-going-to-take-it-anymore." And that's what we end up writing.

The trouble is, mad-as-hell letters rarely produce the desired results. The reader, feeling attacked, responds in kind. He may dig in his heels and refuse to budge on principle. Or he may decide to defend himself vigorously. A ripsnorting relay of letters may ensue. But, unless you were after a regular dose of adrenalin, you won't get what you want very quickly.

So—what do you do when you're mad?

If you're like me, you can't be reasonable and mad at the same time. And you're mad! For me, the only way to start a complaint letter is to write down everything I feel. I take a fresh sheet of paper, roll it furiously into the typewriter, and give my anger and self-righteousness full rein. It feels marvelous! I tell the offending individual what I think of him, his company, his pathetic lack of brains. I heap up insults with glorious abandon. I try to be cleverly nasty. I don't stop until I've expended every ounce of my rage.

Then I'm ready to write my complaint letter.

Does this sound dramatic . . . childish . . . unbusinesslike . . . unnecessary? Well, anger *is* dramatic and childish. It's certainly unbusinesslike. But is seems to be necessary to every human being and, like murder, "it will out." Even if you try to keep it in.

By giving way to my anger, letting it run its course, I make it dissipate. I don't allow it to control my communication—which it would do, were it still buried somewhere inside. Amazingly enough, I can even derive some creative pleasure from it, as I try to write the *most* derogatory, castigating insults ever produced!

I don't throw away all my "mad-rags." I keep some for fun and others to show me how ridiculous I look when I let anger take over. Rereading them, I can imagine how *I* would react if anyone sent me a letter born of those feelings. I would send one right back.

Write your mad-rag—*first*

I strongly suggest that you write your own mad-rag before you attempt a complaint letter. I believe Horace is right when he claims, "Anger is a short madness." And insanity never produced a good letter.

Here's one of my mad-rags to use as an example.

You brainless incompetent,

Paragraph one: unprintable epithets.

I can't believe a company of XXX's stature would hire such a nincompoop as you. Here we are up to our necks in unemployment lines, and people like you are actually drawing regular paychecks. It rocks my mind.

Do you remember, you half-witted turkey, that I talked to you personally *three times last week? I told you YYY had sent me a package for overnight delivery through your firm, at the beginning of the week. Are you aware that people use your "Overnight Delivery Service" precisely because they expect* overnight delivery? *Do you remember that my package had not arrived by Wednesday, Thursday, or even Friday?*

You said magnanimously that you would "put a search on it." I'm ready to put a search on you, you grandiloquent fool, and strangle you with my bare hands.

Oh yes, I got the package today. Only it was the wrong package. My *package contains manuscripts. The one in my hands consists of several pieces of hardware which I would gladly put to use on your witless skull.*

Find my package, Mr. _______, and get it to me by special messenger today, *or I will get the police after you for tampering with the U.S. mail.*

Yours sincerely.

Mad-rags like this one give your anger an outlet. They also show you why you should never write to someone in the heat of rage. Why? Very simply, the subject of your complaint gets lost in the hubbub. You insult your reader—he retaliates—and very soon it becomes a point of honor *not* to cooperate. By the time you both simmer down, the package will be half way to Kalamazoo.

Furthermore, you end up making a lot of rather stupid remarks when you're mad. I couldn't get the police after my bungling private delivery man for tampering with the U.S. mail. There hadn't been any U.S. mail! And just aesthetically—being up to the neck in unemployment lines suggests an image of thousands of Lilliputians eddying around one looking for work. Not a stimulating or valid metaphor.

Writing the mad-rag is important

Perhaps you're thinking, "Yes, I can see the value of not letting anger control my pen. But who has the time to waste writing a letter that'll never be sent? I'd just go outside and tell the idiot what I think of him (in my head) . . . Isn't that just as good?"

No. For a mad-rag offers you a chance to do more than simply burn off your rage. First, it helps you get all your complaints on paper. When your anger's at a peak, you're not likely to forget any of the details that have made you so mad. Later, you might.

Second, the mad-rag gives you a unique chance to see the problem in a more realistic perspective. Rereading it, you visualize the fool who caused you all this trouble, yes. But you also see the letter-writer, bounding up and down in a rage, mixing metaphors, and pulling empty punches. Slowly, both parties begin to look just a little ridiculous. And then—just a little human.

That's when you're ready to settle down and write the complaint letter that will get you results.

We'll see how to do that next time.☐

How to write a complaint letter

Part 2: Seven rules for success

If you've followed the last column's advice to write a "mad-rag," your anger's burned off and you can safely ask yourself the question: "What do I want?" By now the answer won't be: "To string that idiot up by his ears!" Rather, you'll see that you want your problem corrected. And your letter should be focused on getting the reader to correct it for you.

Here are seven rules to help you complain with panache and success.

1. **Think up an agreeable reader.** Imagine that you are sitting down to talk with this person who has caused you so much grief—and that he or she turns out to be a perfectly delightful human being. Forget your preconceptions or bad experiences with him. *Invest* him with qualities he may or may not have. He won't mind your treating him as if he were good, kind, honest, wise. In fact, you may very well awaken those sleeping traits inside him if you write as if they were there!

2. **Speak to your reader as you would have him speak unto you**—if *you'd* make a hash of things. Think of the last time you gave someone legitimate cause to complain. You probably didn't mean to do it. Most likely you were eager to right the unfortunate wrong. Most people are.

Approach your reader with the conviction that he didn't mean to make this mistake—and believe it. Even if he was really "out to get you," you will disarm him by treating him as if he were above such base instincts.

And notice I said, "*Speak* to the reader." Send him a nicely *spoken* letter. That means you could literally say your letter out loud and sound completely natural.

If you write as you speak naturally (not when you're putting on airs!), you will avoid the monstrosities that plague all business letters—inflated, impersonal expressions such as:

We regret to advise that . . .
Upon investigation we find that . . .
This is to inform you that . . .
We anticipate that we will not be in a position to . . . .

Wordy mouthfuls like these beckon to writers of complaint letters. Why?

First, we hope to hide our anger behind the mask of impersonality. If we haven't written any mad-rags, that anger is still there, itching to wield the pen. Stock phrases cover it up. Of course, if you have written a mad-rag, *you're* now in control. You've let your anger speak. Now you can write fearlessly in your own words.

Then there's another attraction. We think great gobs of business language make us sound important. Fearsome, even. We've got the whole weight of the business community behind us, symbolized in those few little clichés. The reader wouldn't dare disobey.

But he would. He's got a million "business letters" in his "In" box. Why should he pay special attention to yours?

Try talking to him, instead. Show human goodwill. It works.

3. **Tell the reader what's in it for him.** Stress the benefits that he will derive by doing as you ask. Don't threaten. Empty threats weaken your argument and encourage the reader to fight, rather than help, you. Instead of saying, "Forget about any more business from us" (even if you couch it in more polite terms), tell him you have enjoyed doing business with him and look forward to placing more orders. Provided this one is filled *pronto*. Don't be afraid to inject a note of humor. Chances are you are writing to someone who didn't cause the trouble all by himself. Your tone of goodwill will assure him that you understand his limited responsibility. But—by showing him that he can benefit by clearing up the mess, you'll give him a powerful reason to do so, *now*.

4. **Don't be sarcastic.** Sarcasm sounds *so* clever on your own lips—and so vicious when it's directed against you. Sarcasm is irony motivated by scorn. It comes from the Greek verb "sarcazo" which means "to tear flesh." That's what it feels like. And it doesn't work.

Look at these two requests:

"I'd appreciate a call on this by the end of the week—that is, if your busy schedule of lunches and golf dates permits."

"Would you give me a call this week to discuss this problem? I'm sure we can work it out simply and effectively."

Which request would encourage you to pick up the phone?

5. **Give a thorough description of the problem.** Impossible as it may seem, you may be part of the cause of the trouble! You may not have given your instructions clearly. You may have omitted certain important details. In this letter, tell your reader *exactly* what you want. Don't behave as if you are talking to a child. But do assume that your reader will look no further than your present letter for details of your request. Be sure they're all there.

6. **Try to give the reader a choice.** It may be impossible for him to fulfill your demands as stated, even though he is at fault. All your ranting and raging won't make it possible. Remember, your object is to *get the problem solved*, not to win points. If your reader is too muddleheaded to sort out all your figures before next month, recognize and accept that fact. You can't change it. Instead, offer him the alternative of sending you half the information now, half in a month's time. Half is better than none! And you will have encouraged that particular reader to comply with your request by showing understanding of his plight.

7. **Be courteous, calm, and firm.** A calm tone establishes authority—yours. Courtesy is essential to any type of communication other than mudslinging. And slinging mud just gets you more mud. But don't confuse courtesy with hesitancy. Your reader needs to know that you will respond with goodwill if he takes care of the problem, but he must also understand that you will act if he does nothing. Decide what you will do if he does not comply. You may or may not want to tell him now; it's really enough that *you* know you will take action. Your sense of resolution will come through.

There are your seven rules for an effective complaint letter. Next time, we'll put them into practice, as we transform a "turn-off" letter of complaint into a pleasant, personal message that gets results.☐

How to write a complaint letter

Part 3: The letter itself

In the last two columns (July, p. 97 and Aug., p. 95), we followed the genesis of a good complaint letter—that is, one that will get your problem corrected. We began with the anger-releasing mad-rag and moved to seven steps for successful complaints. Now let's transform a typical, ineffective complaint letter into one that gets results.

Here's the original, from John Pearson to Samuel Wilkins of the ABC Bank:

Dear Mr. Wilkins,

This is in reference to my recent Master Card bill, charged to my account, dated Jan. 15.

It is to my considerable distress that I discover that the problems I have brought to your attention have not been properly corrected. Although the incorrect charge has been removed, there has been no modification to the finance charge.

I would be most grateful if you would oblige me by attending to this matter at your earliest convenience.

Yours sincerely,
John Pearson

The only blunder this letter does not commit is actual rudeness. It is impersonal. It is certainly "unspoken." And above all, it is hopelessly unclear. The writer does not specify what charge had been incorrectly billed to his account. He doesn't name the item. He doesn't give the date of charge. He doesn't even give his account number. At the end of the letter, we still do not know what the finance charge is for, what charges have been corrected, or whether there is a connection between the finance charge and these mistaken charges.

Finally, after having written three curt, unfriendly paragraphs, John Pearson goes all soft and undemanding. The only reason the reader has for "obliging" Mr. Pearson is to earn his gratitude. Judging from the letter, that will be lukewarm at best.

The tone of the letter suggests that Mr. Pearson is fed up with having to write it and doesn't intend to bother looking up the details. Let the miscreant Wilkins do it—he's the cause of all the trouble.

We may sympathize and share John Pearson's feelings. But neither our sympathy nor his indignation will get him what he wants; a corrected statement, with all finance charges removed.

First, he should write his mad-rag; a full-blown expression of annoyance to get the adrenalin moving and the anger stated and dissipated. Then he can write his letter. Let's do it for him.

Step 1: Think of a nicely spoken letter. Sit down comfortably with the imagined Mr. Wilkins. Paint him in pleasant hues. Make him ready to please but perhaps a little scatterbrained.

Sam Wilkins sits nervously on the edge of his chair, waiting for the ax to strike and preparing to defend himself. Instead, you smile at him and think that he might actually be a good sort.

Steps 2–3: Speak to him as you would have him speak to you, if the positions were reversed. Instead of attacking, start out with praise. His bank must have done something right, or you wouldn't have chosen its credit card. As you start talking, you can combine Steps 2 and 3. You can put Wilkins at ease and treat him with respect (Step 2), and you can show him how he will benefit by cooperating with you (Step 3). For instance:

Dear Mr. Wilkins,

I must tell you how pleased I have been with my ABC Bank Master Card. Your fees are still the lowest around, and I was delighted to note that you have increased my credit limit to $3,000.

Now Samuel Wilkins starts to behave differently. He's neither poised to attack nor anxious to flee. He's expectant. If he attends to you, he may get a special commendation.

Step 4: Remember, no sarcasm.

Step 5: Give him a thorough description of the problem. Forget about what you've told him before. Just explain what's wrong now.

I was glad to see that you have removed the charge for the Eastern Airlines ticket, billed to my account 754-82-9503 on October 15, 1982. As I told you, I cancelled that trip and had requested credit on my Master Card. However, the charge of 895.32 remained on my November 1982 and December 1982 statements. While I waited for the charge to be removed—the finance charges on it kept building up.

Now—my problem is the finance charge. It has not been removed. On the contrary, it has now gone up to $60.20.

Step 6: Now tell Wilkins what you want him to do. Give ------ – choice, if possible.

As you can imagine, I would like this finance charge removed from my credit record as soon as possible. A letter from you confirming its removal would give me great peace of mind. If that is difficult, please do see that it is taken off my next statement. But—I would appreciate a letter.

Step 7: Now the closing paragraph—courteous, calm, but firm.

I realize that computer errors can happen in an operation the size of yours. I'm sure I can depend on you to resolve my problem quickly—and I look forward to receiving a letter from you soon.

And there's your letter. You began by treating Samuel Wilkins as a decent human being worthy of respect. You praised his bank and hinted, by your tone, that you would continue to be a valued customer if he behaved himself. Then you told him clearly and simply what the difficulty was. And you enlisted his help. This letter will neither enrage nor perplex him. He will be eager to do all he can to keep your goodwill and your business.

If you follow this method of trust, goodwill, and clarity—you won't have any more "difficult" letters to write. ☐

Writing under fire

Part 1: Dealing with expected disagreement

It would be nice if communication always happened in a harmonious, agreeable setting, with readers disposed to embrace our views and recommendations, or at least keep an objective point of view. Unfortunately, reality sometimes is a far cry from that idyllic blueprint. There are occasions when most of your readers are opposed to your ideas, ready to pounce on the first wrong phrase, unintended implication, or flaw in your reasoning. How do you approach such a writing situation?

Recognizing a charged communication situation

Your first step is to *assess your readers to determine the likely degree of resistance to your ideas.* What exactly are the consequences for them of accepting your proposals? How will they be affected if others agree with your ideas? Your task in this step is to identify your readers' *potential reasons for disagreement.* Here are some specific points to consider:

- Are some readers likely to take an opposing view because of certain prejudices or experiences?
- Are some of your readers formally or informally associated with a group that would have reason to oppose your view?
- Could what you have to say threaten the security of some readers?
- Does acceptance of your ideas require some readers to make drastic changes in their ways of thinking or acting?
- Does your subject fall into a polarized field, with some readers possibly at the other end of the spectrum from you?
- Could some readers be led to an opposing view simply because they lack certain facts?

If the answer to any of those questions is yes, you're dealing with a situation of potential disagreement.

Setting the right communication goals

Step 2 is to *define realistic goals for your communication.* Generally, you will want to persuade the reader to give up his view and adopt yours. Of course, it's not realistic to think you can change everybody's attitude simply by writing; but it's equally unrealistic to assume you can't change anybody's mind. Setting realistic goals means to write with the right attitude—not one of defiance, despondency, or arrogance. But it also won't do to lose your sense of proportion and launch an artillery attack against a mole hill.

In many cases, it may be unrealistic to expect persuasion from a single written communication—you may need to devise a concerted strategy of support for your position, including a lot of casual conversations, careful probing, preparatory letters or memos, and convincing on-the-scene demonstrations. And, yes, there may be cases when you conclude, after weighing all the facts, that your best course is to do nothing.

Picking communication strategy

In Step 3 you *select your communication strategy on the basis of the previous two steps.* If your goal is to persuade the reader, you must (1) confront the various reasons for disagreement with your ideas and (2) offer some "rewards" (such as continued status as a rational, open-minded person) for joining your side. Let's consider the reasons for disagreement described under Step 1 and see how they might affect your strategy.

Case 1: prejudice or negative experience

Say you're proposing a new promotion approach, and your boss was recently taken to task for two similar schemes that backfired. It's no use trying to gloss over those bad experiences. You will have to discuss them, summarize their causes, and show that your idea differs precisely by lacking those disaster factors. The same holds for prejudices. Acknowledge the opposing beliefs, *never* tear them down—and then go on to show why they don't apply to your proposal.

Case 2: group association

Here you will have to go back to the specific reasons the group would disagree with your view. Then address those reasons as discussed under the other cases.

Case 3: security threat

The best strategy is to minimize the threat. Say you made a study of testing procedures in your lab and want to recommend some simple, rather obvious changes that would reduce testing time by 30%. Your proposal may be threatening to the people who installed the current procedures, as well as those who implicitly approved them by continuing to administer them. (If your solution is so simple and obvious, why didn't it occur to them long ago?) Your strategy is to review, and *play up,* the merits of the current procedures and then show that they can nonetheless be improved by your proposal. In that way you reassure your readers that by accepting your idea they won't in any way discredit their own past actions and beliefs.

Case 4: drastic change in thinking or acting

The main requirement here is to prepare the reader carefully, and generally make the journey into unfamiliar territory as easy as possible—and that doesn't mean you should beat about the bush, but stress logical organization and clear, simple presentation, always showing exactly how to get there from here.

Case 5: polarized field

Fairly review both sides, then show that rational people, *though until now fully justified in holding the opposite belief,* now have no choice but to accept your view. Remember, there is never a place for insults, however thinly disguised, *if* you're writing to persuade, not just to glorify yourself.

Case 6: readers lacking specific facts

This is the easiest situation—basically you just need to provide the missing information before presenting your main arguments. But avoid any intimation that the reader is somehow at fault for his knowledge gap.

Executing communication

With these three steps, you've done the hardest part of the job. What remains is to execute your communication. Some suggestions on that stage will be offered in the next column. □

Writing under fire

Part 2: How to deal with expected disagreement

Last time I discussed the first three steps in preparing to write for readers resistant to your ideas:

1. Assess your readers to identify their potential reasons for disagreement and the likely degree of resistance.
2. Define realistic goals for your communication. Can you expect persuasion from this letter or report alone, or should it be a part of a concerted strategy of support for your position? Or might you have greater success speaking instead of writing?
3. Given your goals, choose a communication strategy that addresses your readers' reasons for disagreement and makes it attractive for them to adopt your view.

Having completed these steps *in writing*, you're ready to execute your communication—that is, organize it and write it.

Organize for persuasion

There is never an excuse for sloppy organization—especially not when you're fishing in unfriendly seas. You will need a detailed outline—the more detailed, the better. Here are the tasks in sequence:

1. Collect your main and supporting points, *using your written communication strategy as a guideline*. The points to be covered almost certainly will include:
   - Your main message,
   - A fair (!) review of the opposing points of view
   - The reasons those views don't apply to your proposal
   - Positive reasons to accept your view.

You must also discuss the weaknesses, if any, of your ideas or findings. Readers inclined to disagree will be quick to find the weak spots in your presentation anyhow, so you'll do much better by admitting them openly. At the least your honesty will win the reader's respect. You can then proceed to show why those weaknesses aren't important or how they can be corrected.

2. Arrange your points in logical order, *keeping in mind your readers' perspective*. (What may seem logical to you might not to a reader lacking certain facts or holding very different assumptions.) If you're discussing a process in which sequence is important, you might organize your points to reflect that time order. Or it may be appropriate to proceed from general to specific, or highest priority to lowest priority. Whatever organizing principle you choose, *stick to it*. In fact, write it down so you won't lose sight of it as you get involved in details.

Always open with a clear, concise summary of your main message. We sometimes labor under the illusion that we can "sneak" our disagreeable main message in between pleasant chitchat and harmless side points. This is a case of wanting to have our cake and eat it too: we want to get our message across, yet we don't want anyone actually to notice it. All we succeed in doing is making the reader feel we're trying to hide something from him—which is the truth. To gain the reader's respect, come right to the point, then backtrack to show the reader why he should agree with you.

Finally, consider where to put the facts, recommendations, or arguments you most want the reader to remember. Psychologists tell us that the best positions are the beginning and the end— material in the middle has a much lower chance of being recalled. This is one reason your opening paragraph should contain the main message and not some social pleasantries.

3. Check the outline draft for completeness and logic of argumentation and presentation. Every item must be tied directly or indirectly to your main point. If it isn't, either it is irrelevant or there are gaps in your reasoning. Are there any concealed assumptions that might be questioned? Have you omitted any areas that may be important to some readers?

4. Now review your revised outline against your communication goals and strategy. Is your approach appropriate for *all* your readers? If not, can you correct it? Or would it be more effective to split the communication into two or more pieces, each strictly tailored to the attitudes of a particular reader group? These questions can usually be answered at this stage, saving you a lot of time and effort.

One more thing: set yourself a time limit for the outline stage. Composing the "perfect" outline can become an excuse for putting off the last, perhaps most painful step: the actual writing.

Put it into words that sway opinion

This is when the urge to doodle or visit the water fountain will be strongest. Resist it. Study your outline. Let it sink in. Then put it on the far edge of your desk, turned so you can hardly read it.

And now *write*—as fast as you can. No corrections, no pauses to search for just the right cliché. Write as you would talk. And don't keep your eyes glued to the outline. Use it as a paragraph prompt, a place to get your next idea.

Walk a mile in the readers' shoes

Now that you have a lively, uncontorted draft, it's time to edit. We'll discuss editing in detail next time. However, two things need your special attention when you're writing under fire:

Avoid overstatements. Exaggerating the facts doesn't sway the critical reader's opinion but only gives him a ready excuse to reject the whole package. Don't write "Failure to adopt this plan will cost us our position as leader in the cardboard market" unless you've shown beyond reasonable doubt that the plan's impact is of that magnitude.

Refrain from emotionally loaded words and phrases such as "silly," "useless," "terrible" (as in "terrible mistake"), "obvious," or "as any halfway intelligent person can see." They don't constitute an argument (although they're usually used in place of one) and only invite the reader to respond in kind—by rejecting your views on an emotional rather than rational basis.

The golden rule in editing is to step back frequently from what you're doing and *view the product from the recipient's side*. If I were X, might I construe this sentence as an insult, or that one as an attempt to blame? When you can make it a habit to ask that kind of question, your enemies will turn into your allies and never know how it happened. ☐

The trip report

Part 1: Preparation

The trip report is one of your most important writing tasks. Think about it for a minute. A trip costs your company a considerable amount of money. It also takes you away from your regular work and so leads to some reshuffling of responsibilities. And what does the firm get for all this expenditure? A trip report! You personally get new knowledge—but you pass it along mainly through that report. The only way most other members of your firm benefit from your trip is by reading what you have written about it.

Now, conferences and other visits are extremely important for keeping a company or a department up to date. Furthermore, these trips benefit the traveler by taking him out of the insular company setting and giving him a chance to see how others operate. But several companies have cut back on employee travel because, as one executive put it, "We don't get a good enough return on our investment! People don't seem to bring back anything much that we need."

The problem is: bad trip reports.

The average trip report—a disaster on paper

The trip report is usually one of the most boring products of the businessperson's pen. The writer rarely considers what the reader wants or needs to know about his trip. Rather, he begins writing with a yawn and a quick trip to the file cabinet to see how other trip reports were written. In line with them, he then turns out a chronological "history" of an event: a list of lecture summaries, a day-by-day report of everything seen at a company, or a lengthy description of a particular machine or operation.

The writer aims to disinterested completeness and adherence to past forts. He doesn't try to make his report readable—after all, he reasons, it's not supposed to be a novel! So, hardly anyone reads it. Some may scan the subheadings to see where the traveler was and what went on there. If they think there may be something in it for them, they put the report aside to read later. Otherwise, it gets a check in the corner and a ticket to the file cabinet—where it will serve as a model for the next trip report that nobody reads!

A good trip report—a boon to business

Bad trip reports are a waste of everyone's time and energy. Good ones, on the other hand, provide some of the most useful information a company can get. Knowing the company's interests and needs, the writer can screen his information to report only that which is useful and relevant. He can also ask pertinent questions on his trip to find out how the new information may apply to his firm's particular needs. A good trip report can benefit everyone enormously—especially the writer, who will probably receive a great deal of praise and many more opportunities to take interesting trips! But a good one absolutely depends on thorough preparation.

How to prepare

You will need to make *three separate sets of notes*. Get yourself *three* notebooks, pads of paper, or sets of index cards, whichever you prefer to use for notetaking. Don't be stingy and try to do it all on the same pad! This little extra effort will save you a great deal of time when you come to write the report.

The first set—answers to questions from home

Well before you leave, try to discuss the trip with your boss and colleagues. Find out if there is anything in particular that they would like to know, and make notes.

When you arrive, look over those notes and determine to attend to them. Keep them in a place where you have to see them *every day*—in your conference folder, in the drawer where you put your wallet at night, or in some such unavoidable area. Then, check each night to see if you've found any answers to the questions your colleagues raised.

When you have some answers, *write them down on the first notepad or set of cards*. Don't try to keep them in your head. If you do, you may forget important details. Furthermore, by writing down what you have discovered, you may find some holes in your information. You can ask your questions while you're on the spot.

The second set—company interest

Whenever you are away, consider yourself the company representative. Get into the habit of asking yourself regularly, "What could my company get out of this?" If there is even a tiny possibility that some new information could be useful or interesting to someone in the firm, *note it down* on the second pad of paper or set of cards. You can figure out its precise relevance later.

The third set—general information

Finally, recognize that you are neither allseeing nor omniscient. Without knowing it, you could see or hear something that is important to someone else in your company. On your third notepad, make some brief notes on all the new information you receive.

When you are ready to leave, you should have the raw material for your trip report all on paper, in three different sets of notes. You will have written down everything that could possibly interest your people at home. Now you must simply organize it into a well-written report! We'll see how to do that next time.

The trip report

Part 2: Writing the draft

In the last column (October, p. 168), I suggested that you collect all your notes for the trip report while you are on the trip. When you're ready to write your draft, you should have three sets of notes in front of you. These notes should show:

1. What your colleagues wanted you to find out
2. What you thought could interest your company
3. Other new information that you picked up on your trip.

I'll assume you've got all your notes in three piles. Now, when you come to write the report, set aside 15–30 minutes in which there are no interruptions. You need that quiet time alone to concentrate and set up the report in the most effective way. This is how you do it.

Find your purpose

Before you write anything, ever, you must know what your purpose in writing it is. *You always have a purpose*—and you can be pretty sure that you'll communicate that purpose to the reader, whether or not you intend to do so.

If your purpose is to "dash off the report as quickly as possible," your careless, unfocused writing will reveal that purpose. If, on the other hand, your purpose is to "impress your boss," you'll tend to write pretentiously, cleverly, or wittily. You may find yourself using words that show off your technical knowledge.

To find your purpose, ask yourself simply: What do I hope to achieve with this report? For instance, the writer of a conference trip report might answer:

A. I want to impress my boss with the excellence of my coverage.

B. I want to persuade my boss to implement a technique explained by one speaker.

C. I want to convince my boss to continue to send me to this yearly conference.

Now, here's a principle to remember forever:

> The closer your purpose is to your reader's interests, the more likely you are to achieve it.

If the technique referred to in "B" is for the company's benefit, the writer will probably achieve his purpose and please his boss with the good, company-oriented work. But both "A" and "C" are purely in the writer's interest. He may be able to fulfill "C" if he writes an interesting, focused report that meets his reader's needs. But "A" is dubious, if a writer's primary aim is to impress his reader with something, all he is likely to communicate is his desire to impress. And, of course, he'll achieve the opposite result.

Know your purpose. Then change it if you don't want to communicate it to your reader.

Begin by telling your reader what he wants to know

Here's another principle to remember forever:

> If you want people to read what you write, fill their needs first.

Glance over your notes. Was there any information that you think would be of *particular* interest to the primary reader of this report? Did you find out something that he or she asked you about earlier? Did you discover anything that could be important to that reader's work? If so, note the topic(s) briefly. Just two or three words pointing you to the right notes will do.

You will open your trip report by telling your reader about the topic of greatest interest to him. Then you'll go to the topic next in importance to him. And so on, until you have covered all the subjects that are of definite interest to your reader.

How does this relate to your purpose? Consider purpose "B". If your purpose is to persuade your reader to try a certain technique, shouldn't you start by telling him about it?

The answer is: First tell him what he wants to know. Then tell him what you want to tell him. If your boss is hoping to learn something else from this conference, tell him about that first. Then he will be far more receptive to your suggestions.

If your purpose is to get your boss to send you on more trips, you can't do better than to fill his needs in your report.

Whatever your purpose, begin your report by answering your reader's questions, meeting his needs.

Then tell the reader anything else you want to tell him

After you've addressed the topics of greatest interest to your reader, tell him anything else you consider important. Explain why you think it is important to the company.

Append all other information

You have told your reader what he wants to know. You have told him what you wanted to tell him. But there is still some information left in your notes which may be useful to someone either now or later on.

Just write: "Other talks are summarized in the Appendix." And give each a paragraph in the Appendix.

Now look back over your draft. Have you followed your reader's needs, in order of their importance to him? Do you think you have achieved your purpose?

To write a useful, readable trip report, then, you should:

- Make the three sets of notes suggested in the last column.
- State your purpose and change it if you don't want it known.
- Address your reader's interests first, in order of their importance to him.
- Tell the reader anything else you want him to know.
- Add other information in an Appendix.

Next time, I'll show you an actual report done this way.

The trip report

Part 3: The report itself

You are a member of Star Company, and you are about to attend the 1987 QXR Management Conference.

Before you go, ask your boss and your colleagues if there is anything they want to know. Make a note of their answers.

The boss says, "Sure! See if anyone has a miracle drug for injecting life into a dying project. You know which one I mean."

Several colleagues remark that they'd like to know what M. Speers, the management expert, says in his keynote address.

Your Raw Material

At the conference, you take three sets of notes: "Boss's and Colleagues' Requests," "What I Found Interesting for Star," and "Anything Else New."

After the conference, you look over your notes. You'll probably turn to the notes on what interested you first! Here's what you read.

"What I Found Interesting for Star"

J. Smith—exciting talk on Japanese management technique called *kaizen*, or continuous quality control. Concept sounds feasible and potentially productive for our group.

Info I picked up on competitors—ABC Company has intensified its xxx marketing campaign—DEF Company is building a new lab at yyy—GHI Corporation is considering acquiring a smaller company in the zzz market.

"Boss's and Colleagues' Requests"

(Boss) B. Andrews—idea for flagging project = a series of informal weekly meetings for project staff—all members can air their ideas, concerns, and problems. Speaker said he'd observed an improvement in project morale and productivity since his company started these meetings.

(Colleagues) M. Speers—good examples to illustrate Speers's pet theme of the importance of long-term research.

"Anything Else New"

Notes on some other new management techniques.

Get Ready to Write

First, put down your *purpose* in writing the report. Here, it's probably: "To persuade my boss to investigate the *kaizen* technique." But—before you go any further, remember the two principles I gave you in the last column:

1. The closer your purpose is to your readers' interests, the more likely you are to achieve it.
2. If you want people to read what you write, fill their needs first.

What are your readers' interests and needs?

Your boss, the primary reader, wants to know how to put some life back into that project. And your colleagues, secondary readers, want to know what Speers had to say.

The ideal, of course, is to relate your purpose to your readers' needs. With this report, you're in luck! As you reflect on both your purpose and your boss's interests, you realize that the *kaizen* techniques of continuous quality control could be used to inject life and productivity into a dying project.

So—your trip report will cover these topics in this order:

1. Primary reader's need = ways to put life into dying project
   a. *kaizen* techniques
   b. Informal weekly meetings
2. Secondary readers' needs = highlights of Speers's talk
3. Other notes of interest = competitors' doings
4. Appendix = all the summaries I made.

The figure shows how such a trip report might look.

TRIP REPORT

These were the highlights of the 1987 QXR Management Conference:

- Two interesting ideas for improving productivity and enthusiasm on a project
- Persuasive examples of the importance of a long-term research policy, by Michael Speers
- Some noteworthy developments in our industry.

Ideas on Project Management

J. Smith presented the Japanese concept of *kaizen*, or continuous quality control. (Describe briefly with an example showing how this might apply to a project in distress.)

B. Andrews said his company had benefitted from informal weekly meetings for project members. (Short description.)

I have appended summaries of these two talks, which I think contain some good ideas that we could explore here.

Speers on Long-term Research

In his keynote address, Michael Speers stated that those industries that had espoused an unbroken policy of long-term research had seen steady growth in the 1980s. (Give examples.) By contrast, the industries that had focused on the short term had dropped behind. Examples demonstrated that this trend was continuing, and Speers urged us to examine our research policy. A summary of his points is in the Appendix.

Developments in Our Industry

(List the discoveries about competitors' activities.)

The Appendix contains summaries of talks and management techniques that could be of interest to Star.

The technical report

Part 1: Getting ready

Question: *How do you write a technical report?*

Answer: *You don't. Not until you are sure:*

- *How important the report is*
- *Who will read it*
- *What questions that reader will want the report to answer.*

These are the three cornerstones of any report. When you can see them clearly, you'll find the question of how to write the report practically answers itself. (Don't worry, though—we'll answer it, too!)

How important is your report? *Very* important! Absolutely, crucially important, no matter whether it's a trip report, a technical memo (really just a shorter, less formal report), or a full-blown quarterly report. That report is nothing less than you and your work on paper. In many cases, it is the only chance you get to tell other people what you have done and why it is worthwhile.

Take a simple trip report. You have been sent, as a representative of your company, to a conference or some other place where you are expected to gather information that may be of benefit to you or other people in your firm. The only means by which you can effectively transmit this information—and the only evidence on which your performance will be judged—is the trip report. You can't go into the president's office and add an important fact that slipped your mind when you were putting the report together. You can't drop a quick memo of "addenda" to your VP-Research—not on a regular basis, anyway. Your information has got to be all there in the report, and it's got to be readable, logically presented, and of interest to the reader.

All right, you say, shifting in your chair: I know it's important. Could you kindly get on to the business of *writing* the thing? Not quite yet. I would like you, the reader of this column, to sit back for a moment and reflect on the last few reports you've written. Have you treated them as signal events of your career? Were you aware, with every word you put down, that you, your position in the company, and the project you were working on were going out for company scrutiny in that thin report? Or did you sometimes stifle a yawn and get down to the boring, routine task of turning out another humdrum paper? I assure you, unless you are convinced of the report's importance, *nobody else will be.* And your report, however neat and complete, will not be a good one. So, we come to:

Rule No. 1. The technical report is of critical IMPORTANCE to me, my work, and my career.

Forget this rule at your peril! It's not a bad idea to write it down and prop it up in front of you to give you a prod or two as you work on the report.

Who will read your report? Most readers of technical reports are either:

1. Technical people—scientists, engineers, other people in research
2. Management—the president, top management, board of directors

You write the report *solely* for these readers. You do not write it to clarify the project to yourself, although that can be a side benefit. Every word, every sentence, and the whole organization of the report should work, actively, to interest the reader. Why? Because that reader is a busy person who is not sitting around waiting for reports. He will choose whether or not to read yours. And he will read it only if it interests him.

Rule No. 2. The technical report must interest the reader and, if possible, answer his questions on the subject.

Get a few 3 x 5 cards. Keep a card for each type of reader. Label each "Management," "Research," and any other category of readers your report may have. In parentheses, write the name of an actual person who will read your report, the name of the president of the company under "Management."

Now, list the questions that reader is likely to bring to your report before opening it. The questions may be general or specific, depending on the subject of the report. For a routine technical report or technical memo, the questions might include the following:

MANAGEMENT (John H. Smith, President)

1. How much has this project cost so far?
2. How much can it be expected to cost?
3. What are the potential profits?
4. When could we begin to see them?
5. Might this work make us more competitive?
6. Could it lead us into a new line?
7. What would be the advantages and disadvantages of getting into that new area?
8. Why should we continue to support and fund this project? (The bottom line!)

RESEARCH (William S. Parks, Research Director)

1. How far has this project progressed?
2. What technical snags has it experienced or overcome?
3. Does it look technologically feasible for us?
4. Is the process or product we are working for here better than those we now have? In what way?
5. Have any new or puzzling facts come to light as a result of this project?

You may want to amend this list to fit the project, its significance, the history of its acceptance in the company, and the character of certain key readers. That's why it's a good idea to have a specific person, or more than one, in mind as you formulate the questions.

Once you have your questions written down, answer them. On paper. *You are not writing the report*—you are doing research for it! Take each question of each reader, look up the data you need to answer it, and write down everything you can think of to answer that question quickly and completely. Pay no attention to your writing style. Jargon—bad grammar—pompous polysyllabics—leave them all. You will correct it for the report. Now, you want to get it down, *in tōto*, and you want your information to flow.

Rule No. 3. Don't try to write the report before writing the report.

This writing is Research. We'll come to Development later—in the next column!

The technical report

Part 2: A pyramid of value

Recently, I prepared a report for a group of researchers who wanted more support for the project they were pursuing. Taking the information they gave me on the project, its purpose, their progress, discoveries, and needs, I wrote the report for them, beginning it with a series of recommendations to top management. These recommendations, drawn directly from the information I had, told the readers why the work was important *to them,* why they would benefit by giving it greater support, and how this support would be used. The recommendations were studied and approved, and the group will receive more funds and more people to work on the project. But—here's the interesting part! Two of the original project members who supplied me with the information for the report exclaimed, "Where did you dig up all those recommendations?" When I told them *they* had given them to me, they just smiled, shook their heads (you can feel a smile and a headshake over the phone!), and said that was news to them.

Why the mystification? The scientists simply hadn't recognized their own wishes and requests *in translation.* From their point of view, the importance of the project spoke for itself. They were prepared to submit a detailed report of their work on the project, feeling that this was enough to demonstrate their need for more money and manpower. However, people in top management had to see why this particular project deserved more time, money, and people—and they had to see it in their terms. My recommendations spelled out the researchers' needs from management's point of view. Seeing them translated into language that spoke of profits, plans, and competition, the scientists did not recognize their own recommendations!

Your technical report is literally a translation—and usually, a multitranslation. Since the readers vary in interest and background, you have to use a different language for each.

You can approach the report as a pyramid, each layer of which has a certain value for certain readers. The amount of space devoted to a "layer" parallels the amount of interest or involvement *that* reader has in the project.

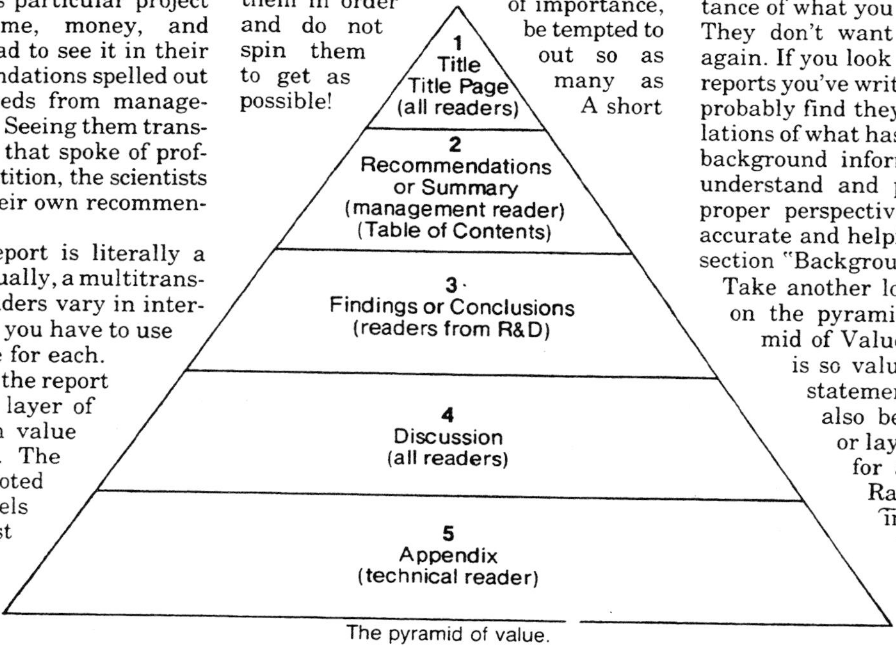

Appendix. At the bottom, layer 5, is all the technical information—charts, tables, diagrams, transcripts of papers or speeches, a selected bibliography—that the technical reader with a great interest in the subject will want to peruse. Layer 5 is also a source of references for anyone anxious to learn more about a certain point. Even though it may be much larger than the actual report, layer 5 is the Appendix—for it includes further information that is not essential to an understanding of the subject.

Discussion. Moving up the pyramid, we come to layer 4, the Discussion. Here, you present all the material all your readers need to answer their questions about the project. (See the last column—*Tappi,* Oct. 1981, p. 131—for a rundown of those possible questions.) You describe the procedure used, any important details of equipment, a comparison of this process with others presently in use, any complications and resolutions you found. But remember, in the Discussion you are giving information, not proving it. The proof for each point you make can be found in the tables you provide in the Appendix. You can make reference to them at the appropriate points in your Discussion.

Conclusions. Layer 3 consists of Findings or Conclusions. These are the discoveries you have made on the project, and they are directed primarily to the reader from R&D. List them in order of importance, and do not be tempted to spin them out so as to get as many as possible! A short page of even three or four solid Conclusions is more readable, meaningful, and impressive than a two-page list of every tiny discovery— many of which have to be dubbed "irrelevant" or "tentative."

Recommendations. Layer 2 is primarily for your management reader. It may be headed Recommendations or Summary, depending on which is appropriate. Here, you extract from the whole body of facts and discoveries the high points of interest to your reader or recommendations for action based on those points. To reach your reader, you really must "put yourself at his (her) desk" and search out the highlights that will capture his attention. Not yours.

Title. Layer 1 is simply the Title and Title Page. The title should tell, in as few words as possible, the *subject* studied in your project and *your specific activity* within this subject's bounds.

You may insert a Table of Contents after the Recommendations or Summary, especially if there are a number of subsections in the parts to follow. If not, it might be wiser (and clearer) simply to append colored index tabs to the appropriate sections.

You may also decide to add an introduction to your Discussion. Approach this with caution! Your Title, Recommendations/Summary, and Findings/Conclusions have already formed a very cogent introduction. Your readers now know quite well the scope and importance of what you are about to present. They don't want to read it all over again. If you look at "Introductions" in reports you've written or received, you'll probably find they are either recapitulations of what has already been said or background information necessary to understand and place the project in proper perspective. Then, it is more accurate and helpful to head your subsection "Background."

Take another look at those sections on the pyramid. I call it a Pyramid of Value because the report is so valuable a record and a statement of your work and also because each section, or layer, has special value for a particular reader. Rather than approaching it as one long development or description, you might consider your report as five sep-

The pyramid of value.

arate translations of a single event. The Title is the most general, sweeping, and comprehensive statement of all. It must be clear to everyone who reads it, and it must tell the subject, the action within that subject, the participants, and their date and place of work. The Recommendations or Summary sketch a broad outline of the significance of the event from management's point of view and in management's language. Findings or Conclusions do the same for the research executive. The Discussion has to incorporate both types of language to tell the story of the event for both managers and researchers, using non-technical language whenever possible but reserving the right to be technically specific when necessary. And the Appendix is really the event telling itself—in documents.

We will look at each level of the pyramid in detail next time.

The technical report

Part 3: Keep it simple

For our final consideration of the technical report—let's write one. Since space is limited, we'll make this a short technical memorandum. A longer report would simply be more detailed, particularly in Sections IV (Discussion) and V (Technical Information and References).

The subject of this report is a new cleaning system you're developing. You are enthusiastic about its possibilities. You would like your readers to be equally enthused—and supportive.

Remember **Rule No. 1:** The report is important. It's as important as your miracle cleaning project. In fact, it can spell life or death for your Mister Clean. Accord the report *time* and *respect*.

Rule No. 2: Make up question cards for the readers of this report. (See *Tappi*, Nov. 1981, p. 124.) Jot down any answers you can give. Now, you can begin your memo.

TITLE. For a technical report, you'd need a full title page giving the subject, your work within it, the participants, and the date and place of work. Here, you have only to fill in the line after *subject*.

A typical title for such a technical memo is: "Recommendations for a new overall cleaning system for X Machine." This is, frankly, dull. Think of your reader's bulging in-box. Now think of the questions he will bring to this report, and try to answer one of them positively. If he read, "*subject*: A new system to increase cleaning efficiency of X Machine by 77 percent"—your memo might receive higher priority. You've shown that your reader will learn something specific from your memo, and you've dangled an attractive plum. Increased efficiency suggests increased productivity.

RECOMMENDATIONS. Here you will recommend support for your new process by showing your readers that *they* will find it worthwhile. Look at your question cards. Management readers in particular will want to know how the company will gain from this process and why that gain is significant. Readers from R & D will want some specifics. So—give them all what they want:

A new cleaning system under study at ST Labs promises to increase the efficiency of X Machine and reduce rejects. As a result, we could expect greater availability and profits. To increase the overall cleaning efficiency of X Machine from the present rate of 0.35 to 0.62, we recommend:

- Installing a larger PQR fan pump on X Machine— . . . GPM instead of the present . . . GPM.
- Using more cleaners of a smaller diameter—×6″ cleaners instead of y12″ cleaners.
- Lowering the headbox consistency from 0.9% to 0.6%.

That's it for *recommendations*. Resist the temptation to swallow them up in a "Summary and Recommendations" section. We can take in only one fact at a time, and you are working to lead your reader through your report as simply and clearly as possible. He will thank you for it.

FINDINGS. This section is a rundown of the evidence on which you base your recommendations. In this technical memo, the *findings* might be as follows:

Our latest studies have shown that the low cleaning efficiency of X Machine can be improved. We calculated the overall efficiency of cleaning systems with (1) fan pumps of different sizes and (2) variations in the process flow. Our calculations showed that, for a given process flow, overall cleaning efficiency rises with:

- increased fan pump capacity
- decreased diameters of cleaners
- decreased headbox consistency

In a full technical report, you could follow this paragraph with some pertinent details of the calculations. This suffices for our memo.

DISCUSSION. The *discussion* on a technical memo can easily cover one to two full pages. We will simply look at the form for this one. The *discussion* is literally an amplification of everything you have said so far. It is also the place to answer any remaining questions your readers might have.

First, you will expand on your recommendations, describing the effects of the larger PQR pump and the smaller cleaners and explaining the meaning and effects of headbox consistency. You can also delineate the various stages of your proposed new cleaning system.

Then, you will explain the calculations used in your tests. Refer to any appended figures or tables that demonstrate your studies or discoveries.

Finally, consider any unanswered questions on your question cards. If you can respond to them, do so here. For instance, management's concerns about money can be answered by the statement that the PQR pump you recommend is not the largest possible but rather the most cost-efficient. Explain this by example, if possible.

In sum, use your *discussion* section to tell your readers everything they want to know about the project: the details of your recommendations, the actual studies and calculations that brought you to them, the meaning of any confusing or technical terms, and the ramifications of this project that may concern these readers. And stop there.

TECHNICAL INFORMATION AND REFERENCES. In a technical memo, you will want to add only the figures you have mentioned in the *discussion*. Such figures might include graphs showing (a) process flows in the present system (b) process flows using a larger number of smaller cleaners (c) effects of headbox consistency on headbox dirt level for different systems, and so on.

The technical memo outlined here looks simple—and it is. It is also rare. Most technical memos and reports are not only complex but baffling. Recommendations and discussions somehow get welded together; technical data of the most abstruse variety find their way into the title itself. If you remember:

1. to write each section for the assumed reader of that section (see *Tappi,* Jan. 1982 p. 60),
2. to keep sections separate,
3. *not* to repeat yourself from one section to another,
4. to answer the readers' questions, if possible,

you will write a simple, readable, and likeable report.

Chapter 5 The Secrets of Communication

How to Make Contact, Listen, and Sometimes Inspire

This section on oral communication begins, appropriately, with lessons from some truly great communicators. A three-part series, **Secrets of the great communicators,** describes how three memorable people made powerful contact with others through passive interaction, empowerment, and commitment. You'll find these skills further delineated in the next two-part series, **Six keys to effective on-the-job communication.**

The first secret of communication is effective listening. This is the part of conversation that most of us do so badly. We define "the listening problem" as the fact that others don't listen sufficiently to us! To correct this misunderstanding and make you a great and rare listener, five columns on **The art of listening** examine listening skills and show you how to apply them to your conversation.

The chapter concludes with several columns that investigate the real message, the one often hidden behind the words we use. **Fake dialogues and "real" messages** and **Think-talk** show you how to recognize your own hidden messages. **Masks** and **The "magic moment" of communication** give you specific instructions for overcoming two of the greatest obstacles to communication. And **Levels of communication,** a four-part series, shows you how to sense and respond to the message beneath the words.

Secrets of the great communicators

Part 1: Passive interaction

The Great Communicators—we have all seen or heard a few. A Great Communicator is one who writes something you just have to read, even though the subject is not your favorite. He is the speaker who holds you under a spell when he is on the podium and makes you like him immediately when you are talking one-to-one.

Over the past few years, I have been studying Great Communicators. These people have turned many of my ideas about communication upside down. They have taught me some truths that I think are essential for anyone who wants to communicate well. In the next few columns, I will be sharing these discoveries with you.

Passive interaction

Recently, I attended a lecture by Patricia St. John, an educator and advanced open-water diver, on communication with dolphins. She told us that dolphins in training would sometimes balk at the trainers' requests for no apparent reason. The more the trainer tried to get them to perform, the stronger their refusals became and the greater the distance they would place between themselves and the trainer.

St. John tried just "hanging around" the dolphins when they were feeling recalcitrant. She did not try to control them; she was just there, waiting passively to see what they wanted to do.

Sooner or later, as she showed us on video clips, the dolphins came to her. They would circle around her and encourage her to join them. Her nondemanding "passive interaction" with them won their trust. For once, here was a human being who was not trying to manage them but simply existing in harmony with them.

In a mind-boggling leap of the imagination, she decided to apply this passive interaction to work with autistic children. Her observations of these children led her to believe that they were suffering from a perceptual overload; their senses simply could not process all the input bombarding them from the world around them. So they "shut down."

St. John just sat quietly with a child, not asking anything of him, not even attempting to engage him in nonverbal communication; she did not intrude in any way. Slowly, we saw the children turn to her. They began to engage her in some interactions. Then she responded—on their terms.

Now this, of course, is exactly what every communicator wants! We want people to turn to us, focus on us, engage with us. But how many of us really listen to understand?

Skill vs. attitude

Before we go any further, here is a warning. This passive interaction is *not* a subtle "listening skill" that you perform to get the other person's attention. If you use it as such, you will fail. In fact, I have come to believe that "skills" is a bad word in matters of human interaction. It suggests that you can manipulate the other person by using your well-honed skills efficiently and effectively. That implies that people are things. And you don't communicate with things; you use them.

What we are really talking about is an *attitude* of deep, fundamental and absolute respect for the other. St. John respected the dolphins. To her, they were not puppets to be jerked into place by a trainer's verbal string. They were independent, complex, worthy beings who had the right to be as they were. By not asking anything of them but simply being there with them, waiting for them to initiate contact, she signified her respect for them as living creatures no less important than herself. Similarly with the autistic children, St. John's silent, passive presence expressed respect, a willingness to meet *on their terms*, and to engage in active communication when *they* were ready.

The first step toward becoming a Great Communicator is to acknowledge that everyone else is as important, as complex, and as feeling as you are. If you reflect on just this fact, "He is as important as I am," before you write or speak to anyone, you will not need listening skills. You will naturally give the other person the time and attention he needs. You will listen to understand him, not to react to him. If he is disturbed, you will just be there, ready to listen, but not to push. You will let the conversation go his way.

As a nation, we spend a great deal of money and time on personal therapy and various forms of psychoanalysis. Ask most people why they go, what they get out of it, and the answer is usually a variant of, "For a whole hour, he *listens* to me." People are crying out to be heard. They value those who listen.

When you have finished reading this, try really listening to the next person you meet. Do not listen to react, listen to understand him. Remind yourself that he is every bit as important and complex as you are and that his thoughts and ideas are as worthy as yours. And then just *listen*.

Secrets of the great communicators

Part 2: Finding the gold

Some years ago, I listened to a talk by a man called Danilo Dolci. He was one of those heroes who "take arms against a sea of troubles, and by opposing" . . . at least oppose them, at least take a human stand. Danilo Dolci led small bands of citizens in Sicily in open protests against the Mafia. Ghandi-like, he marched ahead of the townspeople and taught them to reject fear. He came to Columbia University to talk about his activities, and I went to hear him.

He was electric. We were all on the edges of our seats, marveling at his courage, enlivened by the light in his eyes. After the talk, I joined the line to shake his hand. When I finally reached him, I told him how wonderful I thought he was, how I admired his dedication, fearlessness, and energy.

Signor Dolci turned his shining gaze on me and gave me his full attention, as if I were the only other person in that full-to-bursting auditorium. He said we all had those qualities—or something to that effect. I do not recall his exact words.

What I do remember, and always will remember, is the way he looked at me as if I were someone who could do anything I wanted—anything at all. He seemed to see nothing but good in me, although he knew nothing about me and spent probably a total of two minutes with me. When I said goodbye, I felt empowered—that really is the word—to do anything worthwhile that I wanted to do. He saw a golden person in me who I didn't know was there, and he made me believe that person was my real self. Other people who had talked to him described his effect on them in similar terms.

Danilo Dolci saw the golden person in each of us—and we actually believed ourselves to be what he saw: a little better, a little stronger, a little more worthy than we had thought we were before we met him. Danilo Dolci was a "Great Communicator." He changed the people he met.

The golden secret

Today, if I am feeling defeated or overwhelmed, I can recall vividly the way he looked at me as someone who, just by being human, had limitless potential and resurgent strength. Those two minutes have multiplied into hundreds of hours of hope, energy, and work as I remember his confidence in me and decide that, yes, I can do it, no matter what.

Since meeting Dolci, I have come across one or two people who knew the golden secret. One was a successful businessman, an agent for a book I was writing. The deadline for the book was impossibly tight; with my other commitments, there was no way I could finish it in the time specified. I told him so. He smiled and said to me, "But I know you'll be able to do it. You're like me—you make things happen." I finished that book on time.

The great communicator with the golden secret believes strongly in himself (herself), his work, and his ability to do whatever is required of him. He seems to draw this belief from the mere fact that he is a human being. He does not think he's unique; rather, he believes he can do great things because of certain human qualities that we all share. He has merely had the good sense or good fortune to recognize and use them.

Therefore, he sees these qualities—great reserves of strength, creativity, energy, and brainpower—in other human beings. He knows, perhaps only intuitively, that the way to activate these golden qualities is to concentrate on them and speak to them. So, he looks at them in you, and in so doing, he makes you aware of them in yourself.

When I have tried to apply Dolci's golden secret, the results have been consistently astounding. I become aware of abilities and energy I didn't think I had, and I use them. I feel good about each person with whom I'm talking. I don't see him as a barrier to something I want or a mass of potential objections to my ideas. Rather, I start to see him as someone full of interesting thoughts of his own. I listen more. I look forward to spending some time with this person, or working with him.

Of course, what we say or do depends on the situation, but I can tell you that when I look through the words and actions and roles to see the golden person inside, the person I'm talking to seems to begin to shine. Our time together is vibrant, positive, and often productive. We may not be fighting for freedom together, but our lives and work become more appealing, and we feel like giving more of ourselves.

Try to see the golden person in yourself and others. When you are with someone, believe that one shining human being is talking to another. See what happens, to both of you.

Secrets of the great communicators

Part 3: Commitment

In the past two columns, I've discussed two "secrets" of great communicators: passive interaction with others and seeing the "gold" in others. I've focused on the attitude to the listeners or readers because I believe *attitude*, not masterful use of language or powerful delivery or personal charisma, distinguishes great communicators from others.

A "great communicator" inspires people to change (to try a new method, work harder on a project, or support a proposal). How does he do that? He learns about people by listening to them, really listening (*Tappi Journal*, June 1987, p. 200). He can then speak to their concerns and interests. He encourages them, helps them discover their capabilities by seeing them himself (*Tappi Journal*, July 1987, p. 208). And he motivates them to follow his ideas by *proving his own commitment* to both his message and his listeners.

Knowledge and enthusiasm aren't enough

I once had a geology professor who was a major figure in his field; if it was part of the Earth, he knew about it. Like many other humanities majors, I had signed up for geology because I needed a science course to graduate. I had chosen geology because I thought it exciting to learn about the beginnings and formation of our earth. My interest was philosophical, although I was willing to study the physical details to the best of my ability.

But this professor didn't suffer fools gladly, and, to him, that's what you were if you didn't love geology above all else. He lectured excitedly about details that would fascinate geologists, but he never considered what might be meaningful to the rest of us. He knew his field and was enthusiastic about it, but he was a disastrous communicator. He was not sensitive to his students' interests and did not see anything worth salvaging in those of us who did not share his tectonic obsession. He ruined geology for me forever—to say nothing of my grade average!

I have a friend who writes and speaks beautifully. He has a deep, musical voice and "stage presence." He is well-educated and comes from a famous family. He can be amusing, entertaining, charming, and, initially, persuasive. But this friend's commitment is only to himself and his own advancement. Like the professor, he is not interested in other people. And, although they find him intriguing at first, people soon realize that they don't really matter to him. Despite his good looks, quick mind, mellifluous voice, and effective use of words, he does not manage to persuade others to follow his ideas. In his career, he has not advanced as far as he and many others thought he would go.

Get credibility through honest commitment

I know someone else who does not have either of these men's advantages. Bill is short, stocky, and by no means commanding. He does not have a famous name or a high degree. His words tumble over each other when he gets excited about something, and it is sometimes hard to figure out what he is trying to say. Bill's language and style could certainly use some streamlining, but Bill is a great communicator.

I met him when he was running an educational program that we were considering for our son. We were not at all sure that this was the program we wanted, but Bill convinced us that it was. He listened to our concerns and spoke to each one of them. If he couldn't meet a request, he said so honestly and promised to do what he could to accommodate us. And when he said that, he kept us informed of what he was doing; we didn't have to call him.

Bill asked us what we would like to see in the program. If we made a suggestion that he thought might be beneficial, he did all he could to implement it. He started projects on his own.

Bill was committed to the program and the people—parents and children—involved in it. His commitment showed in his actions: his efforts to try our suggestions, the extra hours he put into finding new projects for the program, the time he always had for us and our child. Bill developed an excellent program and has since advanced rapidly in his career.

Great communicators are committed. If they make promises, they keep them. If they say they will get back to you, they do. They think first about the people to whom they're speaking. They listen to understand, not to react. They believe in the people they're trying to reach. They are generous with their time.

Actions do speak louder than words. The nice thing about that is we all know how to act, so we all really know how to become great communicators—if we only want to!

Six keys to effective on-the-job communication

Part 1

A recent study reported by the IEEE showed that engineers spend a whopping 40-70% of their workday *in conversation.*

With all that talking going on, it may seem strange that people in the technical areas of business regularly complain: "There's no communication around here."

A man comes out of a meeting saying, "After nearly two hours of discussion, I have no idea what any of them really wants." Why? Maybe people were trying to sound impressive rather than be clear. Or maybe no one cared what anyone else meant.

A shipment arrives at a plant in Portland, Ore.—3,000 miles west of its intended destination: Portland, Me. Why? Maybe someone didn't explain fully—or someone didn't listen.

A young engineer goes ahead on a project even though he's not 100% sure of what his boss wants. Why? Maybe he wants to appear to know it all—or maybe he's scared to ask.

The obstacles to effective daily communication are daunting: misunderstandings, false assumptions, lack of knowledge, different meanings—not to mention emotional conflicts, insecurities, anger, fear. . . But rather than horrify you with their infinite variety, I want to give you six ways to overcome them—three in this column and three in the next.

The six keys to communication are easy to understand and difficult to do, for they go against the instincts that make us put up the obstacles in the first place. But if you try them, even one or two of them, the results will spur you to continue. You will start to understand people better, and they will understand you. Annoying mistakes will lessen. Schedules will improve. Honestly! Those who have tried them have found it so.

Key #1: Listen more than you talk

Conversation is two-way: talking and listening. But most people just want to talk: to get their message across, to hear themselves speak, to get their needs heard. Obviously, if each person is either talking or waiting inattentively for the other to stop talking so he can start again, there's no communication. No building of a new understanding.

For communication to take place, someone has to start listening, and you have control over only one person. You. So, make yourself talk less and listen more, to everyone. Once other people notice that you are really listening, they won't need to talk so much. They'll be satisfied that you have, indeed, heard and understood. Then they will start to listen to you. And communication will begin.

Key #2: Remember that *attitude* always comes through

You can't disguise it. If you approach someone with a negative attitude, such as fear, anger, or disrespect, it will show in your tone of voice, gestures, movements, expressions, eye contact or lack of it, pitch, volume, choice of words. And the other person will react to it with equally negative emotions.

So, work on your attitude. Try to approach everyone you talk to with friendly respect. Difficult? It's almost impossible! But it can be done. I don't manage to do it all the time, but I think I'm up to about 80%. And you know, when I do, I find life and work and being with people such great fun. But let me give in to the old resentments or fears, and it all becomes tedious and exhausting again.

Key #3: To bring out the best in people, *see the gold* in them

This is a natural follow-up to Key #2. And, of course, it's even harder. You're dealing with someone selfish, opinionated and lazy. Now I ask you to *see the gold* in him? What gold?

There's a private school in New York City for children with learning and/or emotional problems. Many of its students were expelled from other schools or passed over as "ineducable."

After a year or two at The Child School, these kids can do things no one—including their parents—thought they could ever do. Their behavior changes. Their academic abilities soar. There are many reasons for this, but the core of it is that the director of The Child School and everyone on her staff makes a commitment to see the gold in each child. They call it "the magic"—to see the magic in each child.

What happens is that the child begins to see it in himself. He begins to believe in himself. He begins to try things. A "difficult kid" isn't difficult any more. He's having too much fun doing things he thought he couldn't do.

At The Child School, they empower children to shine by seeing the gold in them.

I know; my son goes there.

That's why I tell you that Key #3 works. See the gold in another, and you'll bring it out. And you'll form a relationship that you never thought possible.

Six keys to effective on-the-job communication

Part 2

In the last column, I offered you the first three keys to effective communication:

- Key No. 1: Listen more than you talk.
- Key No. 2: Remember that attitude always comes through.
- Key No. 3: To bring out the best in people, see the gold in them.

Using these three "keys" will make you an ideal listener. Conversation is, of course, part talking and part listening; the problem is that we all want the other one to do the listening part! If you approach your meetings with people as a listener first and a talker second, you will almost instantly become a top communicator. You'll be one of the very few who actually knows what each party has to say.

I suggest you work on those first three keys for at least another month before you put the next three into practice. If you manage to use Key No. 1—"listen more than you talk"—it alone will distinguish you from practically everyone else in your company. People will seek you out to talk to you, and you will learn more about what they need and what is going on in the firm than you ever did before.

When you have made those first three a part of your daily life, move on to the next. Keys 1–3 dealt with the receptive part of communicating; Keys 4–6 cover the expressive part.

Key No. 4: Put yourself on the same side of the table as the other person, opposite the problem

Besides being a key to effective communication, this is an excellent problem-solving tool. We all tend to complicate situations by confusing the person with the problem. If you have a conflict with someone, ask for time before you go any further. Just say you'll call back or drop by later. Then go to your office, alone. Sit down and quietly identify the problem— without using the other person's name.

Instead of putting it this way: "Joe wants to get all the credit for this project, while I do all the work," phrase it like this: "We both want more recognition and less work." If that doesn't sound right, rework your definition until it sounds like both you and Joe are sitting on one side and the problem is on the other. See the two of you as allies against the problem. Then give Joe a call and go in to discuss it with him.

This communication tool is difficult, because you usually have to use it when you're in a state of high emotion. That is why it is important to go off by yourself, calm down, and then get on with redefining the problem. I can assure you from my own experience and that of others who have tried this: It works.

Key No. 5: Be aware of possible destructive side messages, and aim for positive ones

This one goes back to Key No. 2: Remember that your attitude always comes through. You communicate more through your expressions, gestures, and silences than you do through your words. These "side messages" rarely lie. They tell the other person, usually very clearly, what we think or feel about him.

This key is especially important when you're talking with someone with whom you have difficulties. The way to use it is in two steps, as suggested by the wording of the key. First, be aware of negative side messages. Realize that a sarcastic expression, a "pregnant" silence, a slightly deprecating tone all drown out whatever you're actually saying. Then, aim to give positive side messages.

It won't be easy. But if you remember that your purpose is not to judge the other person but to communicate with him, you will gradually focus on giving positive messages, both verbal and nonverbal.

Key No. 6: Count to five before you react

The last of the six keys is actually the easiest—partly because you will have put yourself through some stiff training with the preceding five, but also because it's a simple act to do.

Someone says something that sparks a response in you. Pause. Count to five while you consider what you have just heard.

This key will serve you in two ways. The few seconds of silence enables you to gain perspective. And your delay in responding will make people take you more seriously.

We all know people who jump in and react without having taken the time to think. We tend to give their words about as much time as they gave ours. If you make that tiny effort to stop, count to five, and then consider your response, your listeners will realize that you have taken a moment to reflect on what they said. Chances are they'll do the same for you! ☐

The art of listening

Part 1: The listening problem

"BORE, n. A person who talks when you wish him to listen." —Ambrose Bierce

In a recent study, the Rand Corp. found committees composed of some of the finest business minds today were producing some of the most useless, unoriginal results in the history of human endeavor. The reason? Rand discovered just one: no one listened to anyone else.

Each committee member was so intent on being heard that he or she never listened to what the others were saying. They spent the whole meeting either speaking or rehearsing their next remarks. There was no time for listening!

The problem goes beyond committee work. The greatest barrier to effective communication—at work and at home—is people don't listen to each other. Listening is half of conversation, yet hardly anyone does it.

Furthermore, most of us don't think of ourselves as poor listeners. We may agree that there's too much talking around and not enough listening, but guess who we think does all the talking and none of the listening. Right. The other guy.

When I teach communication skills, I ask participants to write down what they consider their strengths and weaknesses in everyday interactions with other people. In all my years of teaching this course, no one has ever placed poor listening skills in their weakness category, but several have noted "inability to get others to listen to me." To most of us, "the listening problem" means simply: people don't listen to me.

"No siren did ever so charm the ear of the listener as the listening ear has charmed the soul of the siren."
—Sir Henry Taylor

Human beings crave attention. We value people who give us attention. We need people to listen to us. We even pay them to do it.

A man who met Freud wrote, "Never had I seen such concentrated attention. There was none of that 'piercing, soul-penetrating gaze' business. His eyes were mild and genial. His voice was low and kind. His gestures were few. But the attention he gave me, his appreciation of what I said, even when I said it badly, was extraordinary. You have no idea what it meant to be listened to like that."

When you listen to someone, you're not just absorbing information. You are validating that person. You're giving him a much-needed sense of self-worth. In business, that can be even more important than the transfer of information.

If you want to reach someone, inspire him to work better, or get him to consider your ideas—try listening to him.

Recent studies have shown listening can be improved. In this series of columns, I am going to give you some tested skills you can use to become a first-class listener. We'll look at creative listening, active listening, listening for ideas, the four phases of listening, and more. But, these skills will work for you only on one condition: You must *want* to listen. listen.

Does that sound self-evident and easy? Try it the next time you're talking with someone. Because of our powerful need for attention, most of us don't want to listen at all. We don't want to give attention to the other person; we want that person to give it to us.

Of course, we all listen sometimes. We listen when someone gives us information that we asked for or when a person in power, such as a boss, addresses us. But that's not enough. A good communicator makes a habit of listening to those who are talking to him, no matter who they are or what they are discussing.

For the next month, make an effort to listen more. After a meeting or conversation, go over the other speakers' comments. Did you hear what they said? Did you really listen? Conscious listening must become a habit before you can put other skills to work. If you try them cold, without having built up a habit of listening carefully to people, you will look as if you're shamming. And people resent a fake listener even more than an honest egomaniac. Listening is hard work. It becomes easier when you realize getting attention is not the only enjoyable or useful part of a conversation. By listening, you learn about the subject, the other person, his ideas and fears, and the best way to reach him. You give him a precious sense of self-worth. You improve your relationship with him. And you cause him to respect and listen to you. Is it worth all the effort? According to Charles W. Eliot, former president of Harvard, "There is no mystery about successful business intercourse . . . Exclusive attention to the person who is speaking to you is important. Nothing else is so flattering as that."

The art of listening

Part 2: Three ways to listen

Once, a friend came to me in a rage about a meeting with his boss. Their conversation went something like this:

FRIEND: And now, after all this work, he tells me it's too long. Can you believe it? I told him I was ready to throw in the towel.

ME: (Worried) Oh, you didn't really offend him, did you?

FRIEND: (Turning away) No, as it happens I'm not actually a total idiot. Forget it; I'm sorry I brought it up.

I was taken aback, but I made a mental note to listen quietly from then on and to resist making comments.

But one day, as I sat nodding sympathetically at him, a colleague exploded: "Don't you have any thoughts about it? I though you'd try to help!"

What was I doing wrong? When I began to study listening, I found out. My problem was not failing to listen but listening in the wrong way.

My angry friend needed quiet, supportive listening; my colleague, helpful, creative listening. To listen effectively— that is, to promote the sharing of ideas and goodwill— you must use the right type of listening. There are three main types: supportive, active, and creative.

Supportive listening

The purpose of supportive listening is to help the speaker defuse, calm down, and settle into more rational thinking; the cue to follow for this listening type is an extremely emotional speaker.

Concentrate on giving your complete attention to the person- -not the subject, the person. If you focus on the subject, you'll be tempted to comment or offer advice, exactly what the speaker doesn't want. Erase all other thoughts from you mind and avoid making stray gestures or getting distracted. Think "I'm just going to listen and show him I understand and support him."

An example might be:

SPEAKER: And now, after all this work, he tells me it's too long. Can you believe it? I told him I was about ready to throw in the towel.

LISTENER: I can imagine. You've had a lot to deal with.

SPEAKER: (Calmer) I sure have.

Active listening

The purpose of active listening is to help the speaker clarify a problem. For cues for this type of listening, look for the speaker who talks calmly about a problem but does not ask for advice or involvement. He uses "I" most of the time: "I have to figure out . . .," "I have a problem with . . .," and "I'm not sure how to deal with . . ." Concentrate on what the speaker is saying. When he makes an important point, paraphrase it so as to draw attention to the essence of the problem. Do not add your interpretations or advice; he has not asked for them.

An example of this type of listening is:

SPEAKER: Serena and I have to write the report together, but I'm not at all sure how we're going to do it.

LISTENER: You need both contributions, but you're trying to decide how two people can write a unified report.

SPEAKER: That's it. Exactly.

Creative listening

The purpose of creative listening is to solve a problem with the speaker. Look for the speaker who says he wants to discuss a problem, asks you questions, and uses "we" and "you": "Do you have a moment to discuss . . .," "What do you think we could do about . . .," and "Maybe you could help me figure out . . ." As in active listening, listen for the main points. Restate them or ask questions to make sure they're clear. Then pick up a point and try to add something to it.

An example of this interchange might be:

SPEAKER: He seems interested in the product, but I just can't get him to move. Do you have any ideas?

LISTENER: Do you mean you can't get him to buy? Or is there another step first?

SPEAKER: Well, he'd have to discuss it with his associates first. I don't think he's even done that.

LISTENER: Could you talk to any of the associates directly?

SPEAKER: That's an idea. I know one of them quite well.

Try these new skills. When someone starts talking, make a quiet diagnosis, then use the appropriate skill. If you feel it's not working, switch to another. Remember, if people want advice, they'll probably ask for it. If they don't ask for it, resist the temptation to give it anyway.

The art of listening

Part 3: The four phases of listening

We spend 46 percent of our lives listening.

This is what the Sperry Corporation found in its studies of the four types of communication: reading, writing, speaking, and listening. Yet, Sperry found, although listening is the type most used, it is also the one least taught. We just don't think about it, probably because we don't tend to consider it a willful activity. Speaking is a purposeful act; listening is, well, waiting as politely as possible for the other guy to stop speaking.

Think about the implications of that. We are all standing around trying to give our oh-so-important two-cent's worth; is it any wonder we have communication problems?

Listening—the key to communication

In my opinion, poor listening is the single greatest barrier to effective communication. Good listeners are an absolute—and much valued—rarity. If you learn to listen well, you will be able to:

- Negotiate effectively. You'll be one of the few who really understands what the other side wants, needs, and will accept.
- Deal with crises. You'll be the rare one who *hears* all the different ideas on both the problem and possible solutions.
- Manage well. Studies have shown that the higher you rise in an organization, the harder it is to get people to tell you any bad news. A poor listener is an ignorant manager, but a good listener, one who listens to understand, not to judge, inspires confidence and engourages people to tell him what's going on.

Last month we looked at three basic types of listening: supportive, active, and creative. No matter which type you need to use, the four phases of good listening are the same. You can start practicing them immediately—with the next person who speaks to you.

Phase 1: look at the speaker

If you're not looking at him, you're not listening to him. Your mind is on something else. Be sure to make direct eye contact when he starts speaking and most of the time thereafter.

Phase 2: clear your mind of all intrusions

This is extremely difficult at first. Gradually, though, clearing your mind will become an automatic response when someone talks to you. Common intrusions are:

- Mulling over whatever you were previously doing
- Getting stuck on certain words the speaker uses
- Focusing on his vocal or physical mannerisms
- Letting your mind drift off after a tangential thought
- Reacting emotionally before you've heard it all
- Tuning out because you're sure you'll be bored.

Read over the list a few times, then notice what your own intrusions are and add them to the list. You'll soon recognize and reject an instrusion before it can block your ears.

Phase 3: ask "What is his message?"

Listen for the message behind the words. Try to penetrate through the obstructions the speaker sets up—vague language, emotional overlays, cliches, confused or unfinished thoughts—and keep asking "What's he telling me?"

To probe through to the message, you may have to keep an "internal paraphrase" running through your mind. This means stating to yourself in other words just what the speaker is saying. Internal paraphrasing is a standard listening skill, but it can be confusing and distracting. Do it only if you can't nail down the message by simple attention.

Phase 4: ask "Why is he saying this to me?"

I find this question the most useful of all. It is the fourth phase of effective listening, but it's also the best technique to use if you find yourself wandering away in mid-conversation. It has never failed me; if my mind starts to trot off, merrily following its own trail while the speaker goes on talking to the air, this question snaps me back to attention.

I think the reason it works is twofold. One, this question requires my complete concentration; I'm looking at the whole scene—the speaker, the subject, and the listener—in an effort to understand the dynamics. I cannot do this if part of my attention is elsewhere.

Two, the question delivers so much. It leads me to learn a great deal more about the message, the speaker, and myself. I begin to understand how the speaker sees me, what he wants to share with me, and how he hopes I will respond. When I ask this question mentally, I feel the communication strengthen. I start looking at the speaker as a special human being with whom I have a valuable relationship, and the speaker gets that message. It's probably more valuable than anything I could possibly say, when it's my turn to speak.

The art of listening

Part 4: Listen as you read

Some time ago, as I finished reading a particularly good book, it occurred to me that reading is the best listening I do.

When I read, I concentrate on what the writer is saying. I don't immediately start trying to define or decide what I think about the subject. And, of course, I don't argue with the book while I'm reading it for the simple reason that I know it can't answer back. I might make notes in the margins for later reference, but basically, I read to understand and enjoy the writer's words.

I thought about this. Suppose that writer were to appear before me and express exactly the same ideas that lie printed in his book. Would he find the same interested, receptive listener? I doubted it. I suspected about 25 percent of my attention would be taken up planning my response to his words. Another 25 percent would go on gauging the impression I was making on him. And yet another 25 percent would be spent on how I could improve that impression—perhaps by appearing to be listening more intensely.

A simple act of computation shows that I wouldn't get too much of what he was saying. I wouldn't absorb his ideas, since I'd be too busy figuring out how to make cogent comments on them. And I wouldn't see what he was revealing aobut himself, since I'd be too busy thinking about myself!

I wondered "Could I take my reading skills and apply them to improve my listening?" That is what I tried to do.

Positive anticipation

I began with the way I approach a book or paper. I open it with a feeling of positive anticipation. I'm ready to consider the writer's ideas. I'm eager to say, "Yes, that is an interesting thought." I look for the good, the intriguing, the poetic—and I enjoy doing so.

In fact, I read for fun. I try to get the most pleasure and new knowledge I can out of a piece, no matter what I'm reading. Could I apply this attitude to everyday conversations with people? I couldn't see why not.

Mentally, I noted: Attitude—positive anticipation.

Then I tried it. And got a real shock. It was a tremendous effort! I found that my conversational attitude was negative and critical. As soon as someone made a point, my mind would start bouncing around from one experience to another to try to find an opposing, or at least qualifying, remark.

I made another discovery. Most people bring that negative, critical attitude to discussions. I watched other people talking together; very few listened to the speaker with full attention. Most interrupted with conditions and revisions and tangential experiences. Rarely was anyone able to complete a thought.

I made a third discovery. Positive anticipation brought rewards. When I expected to enjoy the conversation and learn something valuable, I did! I heard more than before. Possibly the people talking to me felt my interest and were moved to talk more fully and deeply about their subject. I believe my attitude of positive anticipation fostered the development and free exchange of ideas.

Total concentration

When I read, I concentrate totally on the text. After all, there are so few distractions. The page is the transmitter, I'm the receiver, and there's nothing else to get in the way.

I tried to apply this total concentration to a discussion. What a challenge! I could make eye contact with the speaker, but I couldn't keep staring at him. Sooner or later, I found myself looking at something else and then thinking about it. Then I discovered that I was focusing on my role in the discussion. I would observe my words and actions critically, as if I were watching an actor onstage. Finally, there were usually several other people involved. It was easy to get sidetracked by them, their relationship with the one talking, and so on.

These distractions made it difficult to sustain total concentration on the speaker—difficult, but not impossible. Indeed, once I adopted the attitude of pleasant anticipation, I had a reason to concentrate on the person talking. I was not about to do myself out of the chance to learn something worthwhile. And, once again, my concentration seemed to encourage people to tell me interesting things. (Or, more likely, I was at last hearing the interesting things people had been saying for years.) Of course, the more intrigued I was, the less I got distracted. Gradually, total concentration became as much of a habit for me in conversation as it was in reading.

Now I listen as I read. I approach a discussion with an attitude of positive anticipation. When someone is talking to me, I give that person total concentration. I pull myself back (and mentally rap my knuckles) if I wander off, but the wanderings have lessened. I'm enjoying the conversation too much to leave!

The art of listening

Part 5: What do you mean by "communicate"?

The voice on the phone said, "I want to sign up for one of your communication courses. I need to learn how to communicate better."

"What do you mean by communicate?" I asked.

After a pause, the speaker said hesitantly, "I guess I want people to listen more when I talk."

The conversation reminded me of one I had recently with an editor of a large publishing firm. Complaining about the accounting department, she said, "You just can't communicate with them. They're on a different wave length. They don't listen."

The first and biggest problem most people have with communication is understanding the concept itself. A glance at the derivation of the word is enlightening.

Communication comes from the Latin word *communicare*, "to particpate." *Communicare*, in turn, comes from *communis*, "common." And *communis* is a combination of *con*, "with," and *munus*, "service." So, "Communication" transmits the idea of *participation*, service for the common good.

Not talking so people will listen.

Communication at its best is like the mixing of two colors to produce a new one. Purple is neither red nor blue nor red and blue; it's an entirely new color formed by the mixture of the two primary hues.

A good discussion is analagous to this "mixing of colors." You go into it with a plan to create or discover something together. You don't focus on getting your point across or using a "power vocabulary" or inveigling the other person into accepting your views. Ideas don't grow this way. People don't work together well this way. If you do approach a discussion with these self-serving motives, the other person will probably do the same, since people tend to respond in the way they are treated. And there will be more stagnation, more frustration, less communication. Instead, make a new move. Think of the color purple—an impossibility if red and blue had not merged. Decide that your purpose is to create something new together, on a common ground.

A common ground

A topical example of communication is Senator Bentsen's handling of the conflict between Governor Dukakis and the Reverend Jesse Jackson.

The two men had engaged in what *The New York Times* called a period of "fruitless name-calling and recrimination," (*NYT*, July 22, 1988). Seeking to reconcile them, Bentsen sought out their "common ground"—a phrase Jackson himself later used in his call for party unity.

"We're facing a tough election," Senator Bentsen told them. "It's going to take all of us to win. Any one of us can pull it down, and all three of us will share in the blame."

Once Dukakis and Jackson saw their common concern—that indeed they did all need each other if they hoped to win the election—they dropped their resentments and proceeded to form a new unity, a triumverate publicly commited to party unity.

Try communication—the original way

Pick out someone with whom you have had difficulty communicating, someone about whom you are tempted to say, "I wish he would *listen* more when I talk!" Now, think about something of importance to both of you. What needs, interests, or goals do you share? Where's the common ground between you? How can you build something good together, something that makes sense to both of you?

Don't plan to use this topic as a way to get him on your side so you can push you point more effectively. It won't work. He will see what you're doing and resent you even more for the pretense.

Instead, note this shared interest or concern now. See it as a new area of communication—literally, a new space where you two will meet to build something different from before, something new. Next time you meet, talk with the other person about this particular common ground.

When you communicate well, you create something: a new idea, a new relationship, a new understanding. Think of the color purple, a new color made by a mixture of red and blue. Somebody has to put down the first color on the common space.

Fake dialogues and "real" messages

The words we use often hide our real message, but the real message is the one that gets understood.

Let's say you are taking a trip with an associate. On the way, you start talking about the problems with the visit—how difficult it will be to see the right people, how much your boss expects from you, how impossible it will be to get the information he wants. Your associate responds soothingly that you should have no trouble getting to see your people, provided you explain your mission to the VP in charge. He says he does not think the boss at home expects more than basic information. He is sure you will have no real problem. Judging from the words alone, you are outlining some problems facing you, and your associate is offering possible solutions to them.

But the dialogue is actually about something totally different. Your associate was the one who originated the idea of making this trip and obtaining the information. However, he will be with you only one day; he departs the next day for a conference, leaving you to get the information and make the subsequent report. You are furious about this. You want to make him feel guilty—guilty enough to cancel his conference plans and stay on with you at the plant where, you strongly believe, his duty lies.

Now, you know your real message is, "You started this. You've got to see me through it." And he hears that message loud and clear. In fact, his remarks are designed not to solve your problems but rather to fend off your attacks. He is rejecting your implied accusations by saying that you are making a big fuss over nothing.

Since both of you know what the argument is really about, what is the problem? There are actually several problems; three of the more obvious ones are as follows:

- **Dishonesty destroys communication.** You and your associate are playing a game, the goal of which has nothing to do with the problems you are ostensibly discussing. You are trying to make him feel guilty. He is trying to make you look unreasonable. Both of you are aware of the other's goals, and both are fighting hard not to be bested. This dialogue is the opposite of communication; communication is a sharing of thoughts and feelings. At the end of a good discussion, participants feel warm about each other, knowing they have given and received something of value. This fake dialogue is a fencing match. At the end of it, no matter how you conclude, you will both feel a little worse about the other.
- **Your plan is doomed from the start.** The one who starts a fake dialogue wants the other to take his statements seriously and respond to them as if they were the real message. Taking the example above, you would hope that your associate would consider the magnitude of the problems you raise and conclude that he must stay to help you deal with them. According to the plan, he would think he had arrived at this conclusion himself. He would have no idea that you had engineered the discussion to lead him there. But, your tone, body language, gestures, and expressions would give you away. The associate would get your real message soon. He would resent your attempts to manipulate him. And he would be angry with you for showing so little trust in him. We do not play these games with people we trust. So, he would fight hard not to give you what you want.
- **The discussion goes nowhere.** Because both speakers are fully concentrated on the real issue, neither takes the subject of the discussion seriously. In our example, neither of you is focused on solving the problems you raised. You are using them to make him feel guilty. And he is just saying whatever he can to reject the guilt. Consequently, this discussion is a total failure. It does not advance understanding. It does not yield solutions. And it certainly does not solve the real problem of the person who started it.

How often do we have dialogues in which the real subject is not the one under discussion? All day long. The "fake dialogue" is probably the greatest source of our communication breakdowns. We do not tell people what we want because we are afraid they won't give it to us. Instead, we try to engineer dialogues that lead the other person to do as we wish without realizing that we've tricked him into doing so. It almost never works.

Try something with me. For the rest of this day, make a mental note whenever you find yourself saying one thing but meaning another. Also note when someone else does this to you. Do the same tomorrow, but then stop the dialogue as soon as you realize it is a fake one. Either state the real message gently and honestly or talk about something else. Do this for at least a week and see what happens to your relationships. Note how people react to you. Watch how your work, and that of your associates, proceeds.

I think you'll be surprised. I'd love to hear how it works for you!

Think-talk

The most powerful messages we give are usually the silent ones.

Imagine that you're in a lecture hall. The speaker enters. Before he says a word, you know what he is thinking about you, the audience, and how he sees himself with respect to you. From his walk, his stance, the expression in his eyes, his hand movements, and many other behaviors, you decide whether he likes you, dislikes you, or is afraid of you. The way you listen to him, the credence you give his words, will depend greatly on the silent messages you picked up from him in those few moments when he entered the room.

Now imagine that you're joining a meeting at your office. One participant is someone with whom you always seem to clash. He looks up, smiles, and says, "Hi." But behind the affable greeting, you hear his thoughts, "Oh, no, now the meeting's going to bog down in trivia and discussions of useless subjects. Why couldn't they have scheduled it for a time this one was out of town?" Again, the person's whole demeanor transmits his thoughts about you to you. And it's to his thoughts, not his words, that you react.

The power-and danger-of think-talk

Think-talk is one of our strongest means of communication. It's also the most frequent cause of communication breakdown, since most people don't know their thoughts are showing.

Furthermore, for all but the most seasoned actors and politicians, think-talk is impossible to hide. We may try to cover up a negative thought with positive vocal or body language, putting on an easy smile or pitching our voice a little lower—but a flicker of the eye, a cough, or a slight gesture can give it all away.

Think about your own unplanned think-talk. When someone difficult comes into the room, don't you feel the volley of predictable, negative thoughts shooting out from you to him? Well, from now on, remind yourself that he hears those thoughts, as clearly as if you'd spoken them. Do you want him to hear them?

If not, it's time to work on "positive" think-talk.

Three steps to positive think-talk

First, recognize that people "hear" what you're thinking about them. (If you're not convinced, you won't do the other steps.)

Second, take half an hour alone, and think about two people who are close to you and two with whom you tend to clash. For each, ask yourself, "What had X said or done that I thought was good?" Write down those actions. Then, next to each action, write down the quality that you think produced it.

This exercise will be hard to do for the two difficult people on your list – but stay with it until you've got at least two commendable actions and top qualities for each. For example, suppose a co-worker whom you dislike heads the local volunteer fire-fighters. Don't dismiss this as "another chance to boss others around." Note it as a good action and write the quality behind it as "Desire to help people in trouble."

You will soon have "quality-profiles" for four important people in your life. Study each profile carefully. Then close your eyes and see that list of qualities next to the image of the person.

Third, when you are with that person, think about those qualities before you say a word. Think, "You're generous, ready for a good laugh, but serious when you have to be." When you do speak, keep thinking about those qualities in the person. Consider them the essence of what he is.

By consistent positive think-talk, you will gradually strengthen your good relationships and take the sting out of the bad ones. Some of the enmities will actually grow into friendships. Just try it.

When you've done this for these four people, try another. And then another. After some time, take a look at your profiles. Are there any qualities that recur or resemble each other? If so, they're probably basic human traits. You can safely assume that at least a majority of any group will have them, too.

Knowing that will help you address people in a more formal situation. Look at an audience and think, "You are thoughtful, honest, openminded" or whatever constellation of traits you derive from your profiles. This positive think-talk will predispose the audience to like you and listen to you. And it will eliminate much of your stage fright, since there's no need to fear decent people. More on that next time. Between now and then, try some positive think-talk. Good luck!

Masks

Most of us put on a mask almost as soon as we wake up. A man may start his day being The Father. Sternly, he commands his son: "Be sure you get to that math homework before you go out today!"

He leaves home and instantly switches roles to The Commuter. Burying himself in his paper, he assumes a closed, unapproachable look, designed to keep as much space around him as possible.

At work, he does another quick change to The Boss. He begins to say and do the things he thinks "The Boss" should say and do. Later, called in to speak with the president, he has to do a fast mask change from The Boss to The Subordinate.

Back home, he picks up The Father or Husband mask again and acts accordingly. If he spends the evening with friends, he'll do another role and mask shift.

Why do we wear them?

Like most pretenses, masks are usually a response to fear. The man in our story is afraid his son and staff won't follow his instructions if he speaks to them as one human being to another. But if he plays The Father and then The Boss, they'll respond as their roles dictate—and they'll do as he says.

So, what's wrong with masks? Very simply, they cause problems that are much greater than the help they seem to bring.

For one thing, they severely limit the thoughts and actions of the people wearing them. Consider a discussion between The Boss and The Subordinate. These masks dictate the possible subjects of discussion (nothing personal), the type of dialogue (subordinate must not interrupt), and the weight given to each opinion (the boss's counts for more). Such constraints can be damaging if the subordinate happens to be the expert.

Thus, by forcing people to suppress ideas, masks hamper creativity and problem solving. They also prevent people from knowing each other. All each knows about the other is how well he wears his mask.

How do we get rid of them?

Masks are a lifelong habit. You won't get rid of them just by deciding you don't want to wear them.

First, you have to recognize them. If you find you are choosing words to make a certain impression or observing yourself and the other person as if you were watching a performance, you can be pretty sure you've got a mask on.

Imagine the mask you're wearing. Exaggerate the features to remind yourself how foolish this game is. If you're being The Boss, visualize your face locked in a tight-lipped frown. If you're playing The Subordinate, see yourself smiling and nodding incessantly.

Then, mentally discard the mask. Smile your own smile. Speak as yourself. And watch the change in the conversation. The other person will start to relax and talk more naturally. Soon, two people will be discussing a real problem, instead of two actors losing themselves in fixed, mindless roles.

How can we get along without them?

Is it really possible to work effectively without a mask? Or do you have to play "The Boss" to get people to listen to you or "The Professional" to convince them of your expertise?

Recently, I gave a writing workshop to a group of people who were clearly afflicted with low morale at work. I asked one of the students about it over lunch.

"It's the new boss," he told me. "He comes down here and tells us what he wants. He's got his routine all planned out; you can feel it. He doesn't want to know what we think. Just follow his orders and that's it."

"And his predecessor was different?" I asked. "Oh! We called him by his first name. He used to come down and talk with us—you know, not AT us. He always wanted to know what we were doing and what we thought about it. Then he'd make suggestions. I enjoyed his visits."

"Did you follow his suggestions?"

"Oh sure. He was the boss, after all. Anyway, he knew what he was doing. He was the best—a first class engineer as well as a good man."

The answer, then, is: Work at being the best. Concentrate on being a first-class professional, manager, parent, friend—and you won't need a mask to simulate authenticity. You won't even want that mask. Your own person will be better than its sham.

The "magic moment" of communication

Imagine that you're sitting at your desk, working. A colleague enters your office. You exchange brief greetings. What happens next?

Usually, the new arrival tells you why he (or she) is there. To put it more bluntly, he tells you what he wants from you. And you have to decide, pretty quickly, how to respond.

The language of business sometimes seems to be rooted in four words: "I want you to. . ."

It's understood that there will be a trade: I want you to give me X (your effort, your approval, your time, a raise, a job, etc.) and in return I'll give you Y (my business, a good word, some help when you need it, etc.).

When one person demands something from another, with a partially hidden carrot or stick to back up the demand, a buy-and-sell transaction takes place. These transactions go on in business all day long. People demand things from one another. If they have something worthwhile to offer in return, they often get their demands met. If not, they usually don't.

This is the language of the marketplace. It's the language of deals.

It's not communication.

Communication vs. transaction

To communicate means "to make common, to share,"from the Latin word *communis*, shared. It does not mean "to make a deal."

No wonder effective communication is such a problem in most places of business. Hardly anyone engages in it.

But everyone longs for it! It's one of the great pleasures of human life. We don't thrive in isolation. We are social animals; we like connection.

Besides being a balm for the human soul, communication is good for business. It promotes cooperation. Connecting through shared thoughts, ideas, feelings, knowledge, and reactions makes us feel good about ourselves and each other. Consequently, we are more willing to work hard together.

However, when we engage continuously in buy-and-sell transactions instead of acts of communication, we feel no connection. On the contrary, we feel more and more isolated. We start to believe people at work don't want to connect; they just want to get something for themselves. Yet all the time, we ourselves are engaging in the same transactions that masquerade as communication.

How do you communicate in a marketplace, where everyone is used to the language of trade? How can you break into a buy-and-sell culture of isolated individuals—and start communicating?

There is a simple way to start. I call it "the magic moment," the first little bit of time you spend with the other person.

The magic moment

If you're initiating a buy-and-sell transaction, you spend the first moment either telling the other person what you want or trying to "prime" him to accept your demands. Either way, that first moment is taken up by an aggressive action: a thrust forward to get what you want.

But if you want to communicate rather than make a deal, try making that first moment one of reception, not action. Your aim is to pick up signals from the other person. What kind of mood is he in? What is he doing? How is he doing? In that first moment of meeting, your attitude tells the other person whether you're there to communicate or to transact a deal.

Imagine, again, that you're working at your desk. A colleague comes in, but he doesn't immediately start talking. Instead, he smiles and takes a few seconds before asking you if you have a moment to discuss something. Or he looks in, greets you, and says he sees you're busy and will catch you later. Now imagine the same person marching in and immediately talking "at" you. Can you feel the difference in your reaction?

If you decide that real communication has value—that it will add to the productivity of your teamwork as well as the quality of your life—you will find you can make a connection through that first moment. Show the other person that you're tossing out a friendly line, seeking to connect, by first focusing on the signals he's sending out. Listen hard. When you feel you know where he is, you can cautiously move to meet him there.

If you think in these terms, you will start communicating. Think of being, initially, a receptor rather than an actor. Think of listening with your whole being to understand the other person's state of mind. You will find you're approaching people in a totally different way.

You will find you have relationships at work—not just dealings. □

Levels of communication

Part 1

A baby willfully drops a cup of milk, then chortles happily as his mother shrieks, fusses around, mops up the milk, and scolds him. That's communication.

A manager walks into his secretary's office and says, "Please have these documents ready by noon." That's communication.

A teenager screams at her best friend, "I hate you!" That's communication.

A mill worker sees a new hire having some trouble on the job. He leaves his own work and goes to help the other man and show him how to do the job more easily and effectively. That's communication.

A hostage arrives home after over a year in captivity. As he walks from the plane to the terminal, he looks around the crowd and his gaze meets that of his wife. That's communication.

A man or woman kneels alone in prayer. And that's communication.

What is the common thread that runs through these and many other very different situations? Why do we recognize them all as examples of *communication*?

The answer, I believe, is that each of these instances shows one person intent on *making a connection*. The essence of communication is just that: reaching out from your personal isolation to make contact, to connect, with someone else.

We do it all day long, in hundreds of different ways. And when we don't manage to make the connection—when the other person reacts in such a way that we know he or she is somehow not thinking along the same lines as we are—we feel frustrated... annoyed...defeated...sometimes even afraid. Then we don't manage to connect, we are each once more isolated within ourselves. We are cut off, alienated. And that is scary. Absolute isolation is probably the closest thing to hell on earth. Solitary confinement is the worst punishment.

Communication, connection, and the Language of Business

What do these musings have to do with the language of business? Everything. *Everything* I have been teaching and writing about communication skills for the past ten years, and the more I study the subject, the more I see that communication problems go back to a single source: *the failure to connect*.

A research scientist writes a project report to management. He spells out all the details, which are interesting and important to him but to management, practically meaningless. He gets little or no response, because he failed to connect.

Two professionals who are thinking of starting a partnership begin to discuss terms. They talk about business strategies, but while Tom is motivated by a desire to create a strong business together, Bill is propelled by anxiety about fair division of the profits. Communication stalls. The two leave the discussion thinking maybe they'll drop the idea of joining forces. They failed to connect.

Toward a better connection

When we speak, write, read, or listen to someone, we are always trying to connect. But to make a connection with another person, we must both travel along the same path. One reason we fail to connect, or lose contact after an initial connection, is that we go off on different tracks.

In the next series of columns, I will identify some of the different levels, or tracks, of communication, from the most elementary level, at which one seeks merely to get the other's attention, to the highest level, at which each person draws on the best in himself to reach the best in the other.

To begin moving toward a better connection, identify your own efforts to connect when you talk or write. By what track are you trying to reach the other person? Are you trying to get him to admire you (Level 1)? To do something for you (Level 2)? Or to build something new with you (Level 3)?

Then, consider your verbal and nonverbal language. Does it reflect the level of your thoughts? Or are you trying to speak on one level when you're thinking on another? To connect, you must bring your thoughts and words onto the same level, or the other person will sense the discrepancy and mistrust you for it.

Now look at the other person. Is he responding on your level, or is he on another track? For instance, as you're trying to persuade him to implement a new idea (Level 2), is he trying to impress you with his own clever thoughts on the subject (Level 1)? To connect, you will need to get on the same track.

Try to recognize these levels or tracks of communication. Next time, we'll look at them more closely and identify the "switches" you can use to bring divergent tracks together. □

Levels of communication

Part 2: Level one—the primal scream

The first level of communication begins for most of us the moment we hit this planet. It's the primal scream: "Hey! Here I am! I need help! Pay attention to me!"

The infant becomes a baby, the baby a child, the child an adult—and with each stage of development, the person builds up more complex levels of communication. But that first level never disappears. The primal scream is an announcement of one's existence. It says simply: "I'm here. I need attention."

At some time or another, we all seek to communicate on that level. I can think of at least two recent occasions when I started to communicate on Level One—or, to put it more bluntly, when I began to put words around a primal scream.

In one case, I had an idea for the basic concept of a joint project; my colleague had another. Now, I was getting worried. I wanted my idea to be used. So, I began to discourse on its merits. On the surface, it was a fairly logical exposition of reasons to adopt my concept—but actually, it was a thinly cloaked primal scream. I was really saying: "Hey! Listen! I've got a great idea! Pay attention to me!"

On the second occasion, I was telling another colleague about a successful after-dinner talk I had just given. I carried on about my main points, my subsidiary points, and the techniques I had used to get them all across. But I couldn't fool even myself for long. All I was really saying was: "Look at me! I did a bang-up job! Pay attention, please." The old primal scream again.

Does either of these situations sound at all familiar? Have you noticed yourself or anyone else caught on this first primitive level? Stop for a moment and think over the times when you talked simply to get attention. I urge you to do this—if not now, when you've finished reading the column—because the primal scream unnoticed can cause some powerful damage.

If I hadn't sensed what was going on in those two instances, I would have run into trouble. If I had gone on bulldozing my colleague with my idea, I would have alienated her and stalled the project. (As it turned out, we developed a concept that was better than either of our original ones). And if I had continued the rhapsody on my marvelous speech, I would soon have annoyed my friend by using his time just to get a few "strokes."

How to deal with the primal scream

It is hard to recognize a primal scream, because you're not feeling reasonable when you utter one. You're tossing around on a wave of panic or pride. But it is precisely this emotional state that can give you the clue.

Start to pay more attention to what you're saying and what you're feeling. If you're very excited about something you've done or afraid about something that may affect you—pause before you speak. Remember that you can never totally cover a primal scream. Something of it always comes through. The other person then gets a mixed message or senses that you're just calling for attention. Either way, it's probably not what you want.

If you suspect you're in the clutches of a primal scream, stop talking. Collect yourself. Either change the subject or ask your listener for his opinion on it. If you concentrate on what he's saying, you'll gradually get the focus off your anxieties.

Listen for other people's primal screams. If you think you hear a muffled one, remind yourself that this is not the time to criticize or even reason with the speaker. He'll only resent you for not filling his needs. He will find all sorts of reasons not to do as you ask—as you would in his place.

Instead, stop everything else and just listen supportively. Show that you're interested. Let the other person talk. Give him the attention he needs. Once he's got it, he'll be able to calm the screamer inside and turn his attention to the conversation. Only then can you start to share ideas.

This is not just abstract advice. I was telling my husband about some difficulties I was having with our son. He made a couple of good suggestions which I roundly rejected. Then he said, "I can imagine how rough it was for you"—and I instantly felt relief. He had given me the sympathetic attention I needed. I could then rise from the primal scream to a higher level of communication, at which I listen as well as talk.□

Levels of communication

Part 3: Level two—negotiation

In the last column, I talked about the first level of communication—the primal scream. The primal scream expresses a thoroughly egocentric state of mind, in which the person sees nothing but himself (herself) and his needs. His words translate a raw, basic message: "I'm here. Help me. I matter."

A surprising amount of communication in business occurs on level one, a thinly disguised or carefully mediated primal scream. Just look back over the past few days at work. Think of things people said to you—the things they asked you to do—the letters and memos they wrote to you. How many were really saying, "I need help, I'm in trouble?" The writer may use a deceptively easy or imperially pompous style to disguise the scream, but the message is still there.

The real problem with communication on level one—aside from its obvious immaturity—is that it rarely works. Most people resent someone who yells, "Drop everything. Attend to my needs." They may tolerate such behavior in a young child—but not in an adult. Even a boss can't get away with it for long. First he will lose his people's respect, and then he'll lose his people.

As we grow older, most of us realize that a primal scream won't work any more. We also discover that life in this world involves working with other people, getting along with other people, even doing things for other people. Gradually, we move up from level one to level two.

From shoe-banging to negotiation

Level two is Negotiation. The speaker reasons: "I need you to do this for me. I don't expect you to do it just because I'm here. So— I'll try to make you feel good about doing it for me."

Most of our conversations, letters, and memos in business take place on level two. This level is still egocentric. The speaker is still in a "me-first" mood: "I need this." He's just using a slightly more sophisticated approach to getting it.

When you do something nice for a colleague—asking about his family or work, writing him a letter of thanks, giving him some new information, letting him have his say before you have yours—see if you're doing it in the hope of his doing something for you someday. If so, you're on level two.

Now, if both communicants are on level two, they get along fairly well. They may not have a soul-satisfying conversation, but they do understand each other. Each puts some effort into keeping the relationship positive.

The problems start when the two are on different levels

Jim has promised Ed a paper for the company newsletter. The deadline is June 1, and Jim is way behind schedule on all his work. Overwhelmed, he thrashes about on level one. In a panic, he tells Ed he needs an extension.

Ed, however, responds on level two: negotiation. He says Jim can have till June 8 but adds that this change will force him to rearrange his own work. Could Jim help him out with that work during the week of June 12?

Ed's response is perfectly reasonable—on level two. But Jim is on level one. He wants help now; he can't think about Ed's problems. He bangs the table and exclaims: "How do I know what I'll be doing on June 12? If you can't give me an extension without exacting your 'pound of flesh' for it—you can put out your newsletter without my paper."

Jim's reaction may seem childish, but it is typical of the kind of clash that occurs, to different degrees, when people try to communicate on different levels.

To avoid such unnecessary conflicts, pause before you react to someone. Try to assess his emotional state and the level on which he is talking to you. Then respond sensibly.

If Ed had been thinking about levels of communication, he would have said to himself, "Jim's not in any state to listen or discuss future events. I'll just tell him I'll extend his deadline. Right now, that's all he can take in."

By recognizing these first two levels of communication, you can begin to:

1. Notice when you're operating on one of them.
2. Decide whether or not you want to continue on that level.
3. Understand the other person's behavior.
4. Know how to keep a positive relationship with him.
5. Try to bring the communication up to a higher, more rewarding, plane. We'll see how to do that next time.□

Levels of communication

Part 4: Level three—working together

The third level of communication is the one that will help you achieve more, create more ideas, and have more rewarding work relationships. Communication at this level brings you something new—something you didn't have before: new thoughts, new ideas, a new understanding of the subject and of the other person.

Level 3: a change in attitude

At Level 1 (*Tappi Journal*, April 1989), you see only your immediate wants—and you utter those wants, loudly. This is the primal scream.

At Level 2 (*Tappi Journal*, May 1989), you still see communication as getting something from the other person. However, you realize that you are more likely to get it if you give him something he wants. This is negotiation.

But at Level 3, you start thinking differently. You stop seeing the other person as someone who can give you something. Instead, you see yourself and him as two people who can achieve or discover something together.

To put it another way—at Level 1, Bob marches up to Jim and demands a ball. At Level 2, he offers Jim a bat in order to get Jim to give him the ball. But at Level 3, Bob goes to Jim and says, "Hey I've got this bat—you've got that ball. What can we do with them?" And at Level 3, the great ballgame is born

Level 3, then, is the creative level of communication. Here, talking and working with people becomes fun. Here, too, you can raise the level of the conversation. By approaching the other person with the desire to build something (a solution, an idea, a good relationship), you can lift the conversation and the situation above the boring flatlands of Levels 1 and 2.

Level 3 brings results

An American company that manufactures fishing gear was crumbling under the combined weight of increased foreign competition and customer dissatisfaction with the product. The president decided everyone would have to take a 10-percent pay cut. The staff, of course, was unhappy—and the technical manager complained to the president in Level 1 (primal scream) behavior. The president went to his technical staff and said, "We've got a problem. Unless we take this pay cut— every one of us, myself included—we're all going to be out of the water. Now, we've got to develop a better product. I'm willing to put my 10 percent toward that. How can we do it? How can we make an A-1 product that will knock our competitors out of the water?"

They sat down together and decided to improve two key components of the product and switch to a distributor that guaranteed faster delivery. The firm recovered.

The president operated on Level 3. He screamed, "Get this stuff in shape, or you're all fired" (Level 1). He didn't try to bargain or negotiate: "If you can get our sales up, your jobs should be secure. . ." Instead, he said, "What can we do—together?" He realized that he and his people could achieve more working together than he could alone.

To communicate on Level 3:

1. Don't think of "getting something" from the other person. That includes friendship, respect, and admiration. Don't look at people as tools or means to an end.

   Take notice of your attitude for the next couple of days. When you're talking with someone, see if, underneath it all, you're hoping to get something out of him. If so—you're on Level 1 or 2, and your communication will not progress beyond the bargaining stage. You may get what you want—but the spark of interaction that can lead to new ideas and a really good working relationship will be missing.
2. Think of building something new—an idea, a solution, a good working relationship—with the other person. Instead of seeing him as a means to an end, think of the two of you as being able to accomplish more together than either could apart. Remember that neither the one with the bat nor the one with the ball can do a lot—but together, they can play baseball.
3. Try to find a common ground between you—a common concern. When you talk with or write to someone, try to find a subject that matters to both of you, or a point of view you share. Then, see what you can build together.

Approach the other person as someone interesting and valuable, someone who has as much to offer as you do. Search for a subject, a problem, or an approach to a problem that is a common ground between you. If you turn "I" and "you" into "we"—you'll find that's real communication.□

Chapter 6
The Daily Dialogue

How to Improve Your Conversations at Work

This chapter begins with techniques for transmitting your message effectively in various situations on the job.

How to preach to the converted...without boring them to sleep and a two-part series, **Getting your ideas across,** both debunk some popular myths of effective communication, replacing them with tested tools of transmission. **Half-questions lead to complete misunderstanding** explains why you should always clarify the reason for your questions. **Excuse me—may I interrupt you?** gives you a way to stop people from breaking in when you're talking. **How to communicate with your staff—with no interference** gives you a technique for ensuring your people get the message you intended them to receive. And **The proper focus** takes away any fears you may have of approaching certain difficult or "exalted" people.

Next comes some extra ammunition for turning difficult situations into successful acts of communication. **Communicating under stress** offers general guidelines for approaching a tense conversation. **When you have to say "no,"** a two-part series, shows you specifically how to deny a request without destroying a relationship. And **Negotiation and communication,** a four-part series, gives you a complete technique for settling differences in a way satisfactory to all concerned.

How to preach to the converted ...without boring them to sleep

A few weeks ago, I listened to an after-dinner speech by an impressive figure. Chief Executive Officer of a company in the top *Fortune* 500, he had traveled widely and had actually created a number of businesses. He was addressing an audience of fairly young businesspeople from a variety of industries, all of whom had attended this function mainly to hear him. Yet I doubt if anyone who didn't take notes could remember a thing the man said.

My notes tell me the speaker talked about the importance of free enterprise in a threatened world. He insisted that American business must not lose confidence in itself. He urged us to remember that we all had a stake in the future of our country's industry and that long-range planning and basic research must not be sacrificed to "quick fixes" and immediate profits.

Everything he said was important. The trouble was—we in the audience *knew* it already. In fact, we probably believed it with as much fervor as the speaker. His eloquent phrases told us nothing new; his sincere urgency kindled few fires.

Most audiences are already converted. Few of us are invited regularly to speak in "enemy" territory. How do you preach to them? You don't. They are already on your side. Instead, you seek to *grab their attention* and *interest* them. And you can, every time, if you follow four simple but crucial rules.

1. Research your topic

This is the most important and most neglected criterion for speaking well on familiar subjects in familiar territory. We tend to assume our audience knows it all and that our job is to produce an eloquent battle-cry to rally the troops. Instead of focusing on facts, we concentrate on improving our delivery, our phraseology, our choice of quotes, perhaps even our flare for the dramatic—all very important, but not enough to make a good speech. Your listeners want to *learn* something from you. Even if you're considered the ultimate authority on the subject, hike thee to the library. Find out everything you can: new facts, remarks by opposing factions, related points that could substantiate or expand your argument. Don't stint on this phase. The success of your speech depends on it.

2. Give them specifics

They *know* the generalities. Yet so many speakers do nothing but restate accepted tenets of the group, using a few well-known facts to support them. No wonder minds wander . . . .

Your listeners want to hear your own specific experiences and discoveries. They want new facts to add to their ammunition and new approaches to bring to a common problem.

Say you're talking on the need for more financial backing of a particular program endorsed by your audience. *Don't* waste time, breath, and histrionics waxing eloquent on the need. They know it. Instead, tell them exactly *why* you want the support . . . what you expect the program to yield if sufficiently funded . . . what similar programs in competing companies or industries are producing . . . how specifically the lack of support may affect your company or industry as a whole. If you're hazy on any one of these points, go back to Rule 1. Look it up!

3. Present the opposition

Do not ignore your opponents, even though there may be none in the room. A complete, compelling discussion of a subject dear to your listeners' hearts demands a consideration of the forces that threaten its success. Take up your opponents' principal objections (three at most), and present them *from your opponents' point of view.* Be the devil's advocate, for a moment. Satisfy yourself that the opposite camp could not possibly express its objections any more clearly or convincingly than you have just done.

4. Take up the challenge—with facts and imagination

Consider your opponents' tenets. Do you possess facts with which to refute them? Then do so, immediately. Can you point to similar events, programs, or policies from other times or places whose results vindicated your beliefs and vitiated the opposing ones? Cite them, giving full chapter and verse. Your listeners will not mind the details; they'll be storing them up for future use!

Take your opponents' arguments that you heartily condemn but can't factually refute. This is your opportunity to use your imagination and a touch of humor. Try carrying those arguments to their ultimate, if astonishing, conclusion.

Suppose the principal opposing argument is, "We can't afford to support this program now because the economy is bad and the project's results uncertain." You might sketch a scenario in which *all* uncertain projects are scratched. Start with Columbus's voyage. Market research would have yielded confusing reports, if any! Go further. What if we only invested time and money in births that we *knew*, without a doubt, would populate the world with productive, intelligent, diligent people? A world without life would result—but, of course, without your opponents' money concerns, too.

So—there's your speech. You begin with thorough research, whose new, positive results you can present along with your own specific experiences and discoveries. You then offer the major opposing views, which you follow with refutations, perhaps slight exaggerations, and a good dose of imagination and humor.

You'll give your listeners what they want and deserve—fresh, pertinent facts to think about, new approaches to ponder, a load of good ammunition for your common cause, and a rattling good speech. □

Getting your ideas across

Part 1: "Three misconceptions"

Between the idea and the reality . . .
Falls the shadow.
—T. S. Eliot, "The Hollow Men"

It can be so frustrating. You have something useful, important, helpful—even perhaps brilliant!—to communicate, but nobody seems to get it. People raise irrelevant objections, agree but don't react, or simply turn down your idea for no good reason. How can you make them *hear* what you've got to say?

You can start by getting rid of three common misconceptions that may be blocking the transmission of your ideas.

Holding the floor*

Many a speaker believes that, so long as he keeps talking and the others keep silent, he has their attention. They're looking at him and nodding their heads intelligently. He's got the floor.

That's probably all he's got. People are *not* that attentive. In fact, every 10 seconds our attention takes a 1- or 2-second break. If we have something on our minds or if the speaker irritates or bores us, our lapses will be longer—30 seconds or more.

That's not all. According to a psychological study, if a person misses one word in a 12-word sentence, he has only a 75% chance of deducing its meaning from the others. And he has only a 50% chance of filling in the exact word that he missed.

And there's more! Your words don't fall onto a clean slate. As soon as you start talking, your listener starts *reacting*. He immediately connects your idea to his own, to his experiences, and to the possible uses and problems he envisages. These associated thoughts crowding into his mind can easily block out your words.

Resist the temptation to hold the floor. Remember that your listeners are not tuned in to you exclusively, no matter how attentive they look. If you want them to hear you, make a few brief remarks and then ask for feedback. It doesn't matter if you haven't gotten your whole message across yet. People digest small quantities better than large ones. Give them one point and then ask specific questions. Don't say, "Are you with me?" Instead, ask, "Do you see why I'm saying this method should be more cost-effective?" Such questions will elicit an attentive response, not a vague nod while the mind's far away.

Resenting interruptions

Ever hear yourself say, "Excuse me! I'm trying to make a point," or "If I could just be allowed to finish a sentence, you might understand what I'm driving at"?

Now put yourself in the position of the listener. How do you feel when someone says that to you?

Resentful. Irritated. Most of all, *frustrated*, because you really needed to interrupt the speaker at that point, to contradict, corroborate, or question what he's just said. Whatever the reason, you could not go on just listening.

If the speaker stops you from interrupting, chances are you'll hold fast both to your remark (to make it later) and to your resentment. And you'll listen half-heartedly, if at all.

Encourage interruptions. They will show you how much your listener has understood and where you may need to modify your ideas. Never demand that your listener be quiet until you finish. He may obey—but you'll be talking to the air.

Attention grabbers

"I've got something very interesting to tell you."

"All right—now let me shock you."

"Listen. Let me tell you something."

You may have used such attention grabbers to be sure you have your listener's attention before expounding on your idea.

They don't work.

Why? Because the listener reacts to the attention grabbers themselves. He immediately tries to figure out what you're up to. He may think, "Sounds like she's going to ask for something: I'd better watch out!" or, "If this is going to be a lecture, I'm going to tune out."

Instead of opening his mind to listen to your idea, he starts using it full strength to determine why you're adopting this urgent attitude and what your urgency may imply for him.

Attention grabbers don't attract attention to your idea; they *distract* attention *from* it.

Don't use them. Open your remarks by being as specific as possible, so your listener can't make the wrong connections and then tune out, thinking he already knows what's coming. Instead of, "I've got a bang-up suggestion for cutting mailing costs here," say, "You know, we could reduce our costs on this mailing by sending the two packages together." Then your listener will react to the *right* message—namely, your idea!

*For an interesting discussion of this misconception, see Jesse S. Nirenberg, "How to Sell Your Ideas," McGraw Hill, New York, 1984.

Getting your ideas across

Part 2: Tools of transmission

Last time (April 1985, p. 139) we looked at three common misconceptions that impede the transmission of ideas: "holding the floor," "resenting interruptions," and "using attention-grabbers." With those out of the way, we can now consider three top tools of transmission.

Setting the stage

Imagine that you're in the theater. The curtain goes up on a living room. A man stands glowering into the fireplace; a woman sits facing away from him, tapping her foot and smoking a cigarette. The stage is set for a confrontation.

Now imagine that the curtain goes up on a burglar skulking around by the fireplace. The man enters from one door, the woman from another. He nods and they both advance quietly on the unwary thief. The stage is set for a partnership against crime!

Here, art mirrors life. We tend to play out the roles into which we are cast. But we don't always realize that *we* can set the stage to foster partnership rather than confrontation.

To get another person to consider your ideas seriously and positively, you must set the stage to make him feel like your partner, not your opposer.

You begin by *feeling* that you are indeed both on the same side. Then you find a common goal that your idea will serve—and you present the idea in relation to that goal.

Say your idea is to hold an informal staff meeting every week, for the purpose of giving everyone a chance to air his or her concerns. Now, you happen to be more "people-oriented" than your boss, who generally counters your human-resources proposals with remarks about productivity and the bottom line.

Don't rush in and tell him you have an idea that you think will improve *the quality of life* for your staff. That's setting the scene for a confrontation. Instead, think how the meetings could serve your boss's concerns—concerns that you share.

You might begin by telling him that you have an idea for a way to improve performance on a particularly troublesome project. Since that's a concern you both have, you'll be setting the scene for a *partnership* approach to a problem. When he shows interest, you then explain how these meetings would enable each staff member to delineate his difficulties with the project in a calm, detailed way—instead of during a crisis on the job.

By introducing your idea through interests you both share, you will set the scene for a joint effort, not a confrontation.

Asking the right questions

You've started by telling your listener how your idea will benefit him. Now you have to lead him to feel the same way.

You can help him to do this by asking him questions. The first is the "why" question.

Suppose, in the above example, that your boss says people wouldn't be able to explain project difficulties when they're not actually on the project, since the problems come up as they work.

Now, you ask him *why* he thinks that's so. Don't ask it to prove he hasn't thought it out and is just objecting because he's ornery. Ask honestly to find out. He will recognize the sincerity of your question and answer it in kind. He may realize his point is not valid—or he may give you something to think about, something that you must incorporate into the final realization of your idea. If that is the case, tell him what you will do.

The second type is the "what if" question. You ask this one to start him thinking along your lines. Say, "*What if* we told each worker to write down the problems he experiences while he's working on this project, to discuss at a trial meeting next Wednesday?" By proposing something specific to consider doing, you take the first step to making your idea a reality.

Agreeing with objections

When your reader or listener raises an objection, resist the temptation to attack it. Remember, your goal is to achieve a partnership with the other person. You want to be on the same side, working together for the same goal. If you're antagonists, there's no way he will accept your idea.

So, try to find something you can agree with in his objection. Suppose he says, "But people won't admit their difficulties in front of the other workers." Then you might respond, "That's true, some of them might be defensive. Perhaps at the meeting we could call on the more assertive ones first, and then move on to the shyer people. Even if they don't feel they can talk, they may have had their own problems resolved by the answers to the other people's remarks. What do you think?"

The point is: strive for a joint effort. Deflect the thrust of objections by agreeing with them and so putting yourself relentlessly on the other person's side. Show your belief and trust in the other person by asking him honest questions and being ready to accommodate them into your idea if they are valid.

I think you'll be amazed at how many of your ideas are suddenly being tried!

Half-questions lead to complete misunderstanding

Strolling into his assistant's office, Richard Sayers asked, "Have you sent that letter to Bob Wilkins yet?"

"Yes, yes I did. This morning," the young man replied.

"Oh. All right, very good."

Richard went back to his office, annoyed with himself for not having thought earlier about including the latest test figures in the letter to Wilkins. Now it was too late, since Mike had already sent the letter. But it was a pity; those figures might have given his cause the boost it needed.

While Richard was mulling over his mistake, his assistant, Mike Lewis, was having a mild attack. He had totally forgotten about the letter, which Sayers had asked him to send yesterday or, at the latest, today. Frantically, he scribbled out a hasty draft and scurried off to his secretary with it, begging for immediate dispatch. He then stumbled back to his office, exhausted. He hoped he'd told Wilkins everything he was supposed to say; if not, it was too late now.

Clearly, this unfortunate situation could have been avoided if the communication had been better. But—what exactly was the problem in communication?

Half-questions confuse

You might say, "The problem was simply that Mike Lewis lied. If he'd stated plainly that he hadn't sent the letter, he could have written it in peace and included the set of figures."

True. But Mike's response was only part of the problem. Mike was reacting to what he thought Richard was saying. The message Mike got from Richard was: "I'm checking to see if you sent that letter I told you to get off by this morning." Not knowing Richard's motive for asking, Mike supplied the most likely one *and then reacted to it.* He lied not in response to Richard's question but in response to the message he thought was implied.

The communication problem began when Richard asked *a half-question.*

Half-questions are questions that don't show your reason for asking. Very often, a half-question leads your listener to react to a message you never meant to send.

To transmit his whole message, Richard could have said: "If you haven't sent the letter to Bob Wilkins yet, I'd like to include the latest test figures in it." Then, Mike wouldn't have felt threatened. He would have had no reason to lie.

Asking a whole question is not as simple as it may sound, because we generally cut off part of our messages for good reasons. In this case, Richard knew his assistant was a nervous worrier. If Mike had already sent the letter, why bother him about figures it was too late to send? Wanting to spare Mike any unnecessary concern, Richard decided not to mention the figures unless Mike hadn't sent the letter.

Avoid half-questions

However kind your motives for cutting off parts of your message may be—resist them. Half-questions often lead to complete misunderstandings, because *they usually hide the reason you're asking the question.* That's the half you tend to lop off, for reasons such as Richard's. As a result, your listener supplies the reason himself—frequently the wrong one.

If you sometimes feel compelled to follow a question with something like, "The reason I'm asking is . . ."—you're probably asking half-questions. From now on, plan to give your reasons at the time you ask a question, especially one whose motives might be misinterpreted as threatening. A good way to do this is to use an "If . . . then" statement instead of the simple questions:

"*If* you haven't started that job yet, *(then)* I'd like to make a few changes in the introductory phase."

Instead of: "Have you started that job yet?"

Don't react to messages that may not exist

Make a point to notice half-questions other people ask you. And catch yourself *before* you react to a motive that may not be there.

Suppose your boss asks you: "How much time did you spend on this report?" *Don't* immediately conclude: "He thinks I did a lousy job!" If you find yourself thinking such thoughts, put them firmly aside. You may be right—but you may just as well be wrong.

Instead, you can try to use your answer to ascertain the reason behind the questions:

"It's difficult to say, really, when you consider the planning time and the hours spent collecting other people's input. What did you mean, exactly?"

This may prod the speaker to elaborate. If he doesn't, or if the situation or relationship precludes any such detective work, just try to answer the question as clearly and simply as possible:

"Roughly 12–15 hours. That includes the time I spent planning the report and collecting other people's information for it."

Half-questions lead to misinformation and a lack of communication. Avoid them yourself, and resist attributing motives when someone else asks them. Your communication will improve from all sides.

"Excuse me—may I interrupt you?"

Do you *feel* interrupted? Does it seem to you that you never get to finish a sentence or a thought because somebody always breaks in? And does it make you mad?

If so, you may be able to avoid a great deal of aggravation by correcting one or two mistakes.

1. You may be unwittingly giving signals that you have finished talking—when it fact you haven't.
2. You may be doing something that consistently induces other people to break in.

Let's consider the first prossibility first.

Faulty signals

If you're giving signs that you're ready to turn the conversation over to another speaker when you're not ready at all—you're in good company. A recent study demonstrates that British Prime Minister Margaret Thatcher is interrupted more frequently than other senior politicians because "she displays turn-yielding cues at points where she has not completed her turn."* The "turn-yielding cues" disclosed in the study are:

A drop in pitch and volume. When you let your voice fall, your interlocutor assumes you have finished saying what you wanted to say. Listen to a conversation. You will find that each speaker's voice usually "goes down" when he's finished talking, i.e., his last words are uttered more softly and at a lower pitch. Since this drop in pitch and loudness seems to be a universal signal that one has finished talking, the speaker should not be surprised if someone else starts. He's not trying to interrupt; he's just taking his turn. Try recording your own conversation one day to see if you are giving this faulty "turn-yielding cue" before you're ready to give up your turn.

Ending a gesture. In conversation, we all make gestures to amplify or explain what we're trying to communicate. Depending on your personality and provenance, the gesture may range from a sweeping full-arm flourish to a slightly raised eyebrow. It's still a gesture—and it has a beginning and an end. If you *conclude* the gesture—by dropping your hand, lowering your eyebrow to its normal position—you also signal that you have said your piece. Be aware of your gestures next time you talk with someone. Are you signifying an end when you're in fact still in the middle of a thought?

Looking someone straight in the eye. Although good, honest people tend to make eye contact during conversation, few of us fix our associates with a steely stare from start to finish. We usually begin talking directly to the person and then let our gaze wander slightly as we search for the examples or expressions we want to use. When we turn back to the person addressed and look him or her straight in the eye—that direct gaze is interpreted as a signal that we have finished talking and are waiting for a response. Are you giving crossed signals?

By learning to control these three major signals, you can avoid giving mistaken impressions of conclusion in your conversation. But what if you still feel interrupted? Then it is time to consider possibility number two.

Inducing interruptions

From examining myriad conversations and broadcast interviews, I have found that people are moved to interrupt a speaker when he:

- repeats himself, in the same or different words, to make sure he has been understood
- aims at completeness, giving every detail he knows about the subject whether or not it directly affects the topic under discussion
- tells the other person what that other person believes or is thinking
- makes an assertion aggressively
- talks for more than two minutes

Do you see the pattern that emerges? People are interrupted *when they mistake their role in the conversation.* A conversation is not a solo performance. Rather, it is like a tennis match. Its existence depends on sending the ball back and forth. If one player dances around indefinitely on his side of the net, bouncing the ball cleverly on his racket, the game stops and the other player naturally wants to get it going again. He urges his partner to send the ball back.

The person who gets interrupted is very often the one who has forgotten the rules of the game. He has taken on the role of speaker-before-an-audience, solo player—star performer. He assumes, perhaps unconsciously, that the conversation is being held so that *his thoughts may be heard.* That is really all he cares about. He is not conversing—he is making a speech. But the other participants don't accept the role of "audience." That's not what they came for. So—they interrupt to get the ball moving again.

Think about it for a minute. If you're interested in an *exchange* of ideas about a subject (Webster's definition of a conversation), you will want to listen as much as talk. You will say only what is necessary to contribute to an understanding of that subject. You will *not* want to:

repeat yourself · be exhaustingly exhaustive · tell the others what they're thinking · be clever, aggressive, or longwinded

And you will have eliminated the reasons for interruption.

Furthermore, if you've taken the focus off yourself as speaker and put it back on the subject, where it belongs—you won't mind so much if someone does break in. You probably won't even notice it. ☐

---

*Beatty, G. W., Cutler, A., and Pearson, M., "Why is Mrs. Thatcher interrupted so often?" *Nature* 300: 744(1982).

How to communicate with your staff— with no interference

Communication occurs when one person talks or writes to another -- *and the intended message is the one the listener hears.* This happens very rarely. Interference between speaker and listener usually deflects and confuses the message, though both people think communication has taken place.

Interference can be *the speaker's unspoken signals:* the expression in his eyes . . . his tone of voice . . . even the words he chooses.

It can be *the listener's expectations.* If he expects the speaker to say or mean certain things, he'll tend to hear those things, no matter what the speaker is actually saying.

And it can be the difference in basic assumptions. The speaker may assume the listener will do A; the listener thinks his job is to do B. When the speaker upbraids his listener for *not* doing A — the listener feels so outraged at this unjustified attack that he doesn't even hear what the speaker is saying.

Interference at work

One day, in a certain advertising agency in New York City, a high-powered account supervisor breezed into his new assistant's small office.

"Call these stations and see when they're going to run our 30-second spot," he purred. He patted the young assistant on the shoulder and shimmered out.

First, the young man passed through a series of hot and cold flushes. This was his first assignment from the supervisor — and it looked like an important one.

But he was confused. This supervisor was known for his dynamic personality and his success at getting what he wanted. What did he want here? He thought back to the supervisor's brief visit. The man had indicated confidence in him by speaking so casually and patting him on the shoulder. That meant he figured the assistant would understand what he *really* wanted. He wanted the job done the way he himself would do it. With panache, persuasion, success.

The assistant drew himself up tall. This was a big one. His job was obviously to get those stations to run the 30-minute spot at prime time. And he would do it.

He settled down to a full day of telephone work. He called stations all across the country. He demanded to know when the commercial was to be run and immediately checked the time against his list of prime times. If the two didn't dovetail, he pleaded, cajoled, demanded, even threatened, depending on his listener's telephone personality. Several hours later, he hung up in pleased exhaustion. He felt he had made some waves. And he had.

The supervisor's phone didn't stop ringing. Outraged voices suggested that he had perhaps lost his mind. What did he think he was doing, having insolent young men telling them what to do with their time? Was he under the illusion that he, personally, was now running the broadcasting industry?

The supervisor stormed into the assistant's cubicle. He was out to get the young man fired, on the spot. When that unfortunate gasped that he was only doing what he had been told to do, the supervisor was both furious and dumbfounded.

"I told you to *find out* what they had for us — not tell them!" he yelled.

But we know what passed through the young man's head. First, he misinterpreted the supervisor's casual nonverbal signals. Then, he expected someone of this man's repute to want more than his words conveyed. Finally, he assumed his job was to get the best time spots he could. His eagerness to do well nearly cost him his job.

Yet his mistaken course was really not his fault. The supervisor didn't give him a chance to check his assumptions. His lordly, if casual manner would have frightened the young man from asking a "dumb question" anyway.

If the account supervisor had tried to communicate, he would have begun by putting himself in the assistant's position. He had been new at the game himself, once. He would have realized that a full explanation of the assignment might be necessary.

Then, the supervisor would have asked himself what the new assistant needed to know. And he would have told the young man that he was to call the stations to be sure that certain ads were scheduled to appear. He would have explained that he needed to know whether indeed the stations were ready to air these ads in the next few weeks. If the answer was "no," the agency would go elsewhere. The call was simply to find out.

He would have told the assistant when to complete the calls and how to present the information to him.

And then he would have asked him, sincerely, if he had any questions.

He would have found a zealous young worker, eager to do his bidding and do it well.

Four steps to communication

Before you speak to a member of your staff, be aware of your visual and tonal messages. Put yourself in control of your side of the communication by determining not to send confusing suggestions. Then, realize that your employees expect you to say or mean certain things *because you usually do.* If you have something unusual to say, *tell* them so. By preparing them to expect the unexpected, you open their ears.

Finally, when you give your instructions, do so in four steps:

- Put yourself in your staff member's position.
- Tell him what he needs to know.
- Let him understand what you expect of him.
- Encourage him to ask for explanations of anything he finds unclear.

These four simple steps will get your *intended* message across — with no interference.□

Communicating under stress

Bill is about to present a report on his group's work to the president and other members of top management. Jim has just been promoted, and he's preparing to hold his first meeting with his new staff. Rachel's boss has just asked her to come in and give him a quick rundown on an idea she's been refining for weeks. These people have one thing in common: stage fright!

The meetings they're facing are not uncommon, especially for active, ambitious business people. But that doesn't mean they're not petrifying. Imagine—in a single encounter, you have to convince other people that you know what you're talking about, and that it's worth their time to listen! It's exactly like preparing to walk out onstage and immediately capture the attention of an unknown audience. If you're not superhuman, you're scared stiff.

All actors admit to stage fright. But the good ones know how to control it. The techniques they use are not limited to the theater. They work equally well in the conference room or the boss's office. Here are two that will help you quiet your fears and reach your "audience"—no matter how formidable it may be.

Assume the role of authority

Before you go into any interview or meeting, take a moment alone. Then make these three statements quietly to yourself:

1. I know enough to present this well.
2. I believe in myself.
3. I believe in them (or him or her).

This is not theatrical witchcraft; it's basic psychology. You're using your conscious mind to calm your subconscious, which is at the moment sending out nervous tremors. You're replacing irrational fear with a sense of authority based on facts.

Statement 1 *is* true. If you didn't know enough, you wouldn't be in a position to talk to these people in the first place. Remind yourself that your message is important and that you are the best one to deliver it. As you start thinking about your subject, you'll stop worrying about yourself.

Say Statement 2 out loud. Repeat it over and over until you really believe it. You're using the conscious mind again, to convince the subconscious that the person it's running is a winner! Once you've convinced it, you will feel your confidence grow.

Do the same with Statement 3. Tell yourself firmly that the people to whom you are about to speak are good, honest, and fair. In so doing, you will make them respond well to you.

Psychologists have found that our response to other people depends greatly on *their* perception of *us*. If someone sees you as honest and good, you are likely to react in two ways. First, you will *feel* honest and good, and will behave decently to him. Second, you will tend to think of *the other person* as honest and good—and will treat him as such.

If, on the other hand, that person sees you as untrustworthy, you will consider him untrustworthy.

Trust and respect the people you are about to meet, and you can expect trust and respect from them, too.

Repeat the three statements two or three times, by yourself, until you feel calm and positive about the meeting. It won't take long.

Be deliberate

Quick, truncated speech and movements communicate nervousness. Even if you have achieved a confident calm, you may still have a habit of talking fast or moving jerkily. Your voice and gestures could then express a tension you don't feel. And your listeners will see you as nervous and unsure of your material. Or—you may still be a bit nervous! Either way, deliberate speech and movement will help you settle any remaining jitters and will communicate poise and control.

From now on, cultivate deliberation. You can do this easily in three steps:

1. **Speak more slowly and evenly in everyday conversations.** If possible, make a tape recording of yourself talking at home or with a friend. If you actually hear yourself talking in a hurried or staccato way, you'll be more determined to slow down!

2. **Finish every gesture you make.** If you start to raise your hand, don't drop it midway. Raise it, and bring it back deliberately. Imagine that you are tracing a complete shape in the air.

3. **Replace nervous movements with expressions of calm.** If you find yourself drumming your fingers or tapping your foot, take a deep breath and smile. If the nervous jitters return, turn them into a small gesture that traces a whole shape.

When we're afraid, our thoughts and gestures tend to go out of control. These techniques will enable you to take charge of them and direct them to express confidence, discipline, and calm. In so doing, you will replace wild fears of failure with a quiet determination to communicate your message. And your audience will listen.

The proper focus

A professional writer remarked that, when he was introduced to people as a writing consultant, they rarely responded with a simple, "Hello." Instead, they balked, smiled weakly, and said, "I'd better watch my grammar."

A recent international study showed that, of all the peoples examined, Americans were the most self-conscious about using their native tongue.

Our biggest problem with language, it seems, is fear. We're afraid of what the language we use says about us. If we misuse or misspell a word, we may appear uneducated. If we mistake the meaning of a technical term, we may appear ignorant. If we mix our metaphors, we may appear foolish. Finally, if we don't use enough long, vague, impersonal words, we may appear unbusinesslike.

What can we do to get rid of this tongue-tying fear ?

Imagine a telescope . . .

To give you an answer, I must ask you to indulge me in a metaphor. Imagine that you have a telescope. You set it up to scan the heavens—and then you step back to look at your instrument. It's rather old. It doesn't have the latest lenses. The lenses it does have are a bit dull. One even has a tiny crack in it. As you look harder and harder at your telescope, you become more and more appalled. How can you hope to see the stars through a piece of cracked and muddy glass?

You clean it for a while and try to adjust it so that the crack doesn't get in the way. No good. Every time you correct one part, you realize something else is off. You're ready to pack the thing up and stash it back in the basement, when a fellow astronomer suddenly dashes up to the roof beside you. He's yelling that there are all sorts of wonderful multicolored dust storms going on on Venus. You rush back to your telescope, adjust it as best you can, and concentrate on Venus.

Do you know what happens? The crack disappears! You don't see it because you're too busy focusing on the planet beyond the lens.

Of course, you'll still try to replace the cracked lens and clean the others during the day, when you're not using the telescope. But while there are interesting planets to be viewed, your imperfect lenses are better than none at all.

Shift the focus from the instrument to the object

The parallel is clear, of course. If you concentrate on your listener or reader, you can't focus on your own linguistic deficiencies. You can—and, I hope, will—eliminate them at the appropriate time. That time is the period you set aside precisely to focus on the grammatical and semantic constructions that confuse you. The time to correct them is then.

To put it another way: We can focus on only one point at a time. That's the way we're made. If you're focusing on your listener, you can't focus on your language. And if you take your focus off your language, you won't be afraid of it!

To be free of fears—look away from them

If you want to get rid of the fears that make it hard to speak or write, shift your focus from you and your words on to your listener and his needs.

Remember, we talk and write to reach the other person and influence him to act on our words. Our success in this endeavor depends not on how eloquently we speak but on how well we find and then fill that person's needs. And the way to do that is to focus on the other person and try to respond to his needs, his questions, his concerns.

This principle is absolutely true, and it works! But it works only when you take it out of the abstract and apply it to your daily writing and talking. So, I would like everyone who is reading this column to undertake an assignment. Please do it today or tomorrow. After that—you'll find excuses and drift back into your fears!

Assignment

Go up to or call someone with whom you have always had difficulty talking. Ask that person about something you have both discussed before, something you know concerns him or her greatly. Determine to find out:

1. How the situation has changed since you last talked to the person

2. Exactly what the person's present concerns about it are.

When you have finished the conversation, write down what you discovered—how the situation has changed and what the other person's latest concerns about it are. If you can't write down this information, you didn't do the assignment correctly. You weren't focusing completely on what your interlocutor was telling you. Call someone else and try it again.

Later, as you reflect on this assignment, you will see that you had little trouble talking to the other person when you were trying to understand him. But as soon as your focus went back onto the act of communication itself—the fact that you were talking to him—you probably became nervous and somewhat lost for words.

When you find yourself tied up in words or bereft of them altogether, think of the telescope. Imagine yourself moving in on it, looking through it instead of at it, and trying to understand the person at the other end of it. You'll be amazed!

When you have to say "no" ...

Part 1

The young man sat facing the man behind the desk.

"I put all the work into this project," he protested. "I arranged for outside help; I coordinated everyone's contributions; I even worked overtime. It's *my* project. So why should Bill go with you to present it to the client? All he did was sign the report!"

This was the third time his boss had listened to a variation on this theme. He had called the young man in to explain why he couldn't join them in the presentation. But the fellow was still carping about it. The man behind the desk felt his patience snap.

"Look, I've tried to tell you why that's impossible. Bill is a manager; you're a technician, period. As you don't seem to understand the dynamics involved, I'm afraid there's nothing to be gained by prolonging this conversation. We both have serious work to do. Good day to you."

Clenching his fists, the young man rose and left. Two weeks later, he announced that he had found another job. His boss was incredulous.

"All because I didn't take him along to make the presentation?" he wondered. "It's not possible."

But it happened. This scene, in fact, occurs daily in offices everywhere. The reaction of the aggrieved worker depends on his character and the importance of the request to him. He may be moody for a few days: he may resign.

Can you say "no" successfully?

Yes. But you may have to modify your definition of success.

When we have to disappoint someone—and that is precisely what we are doing when we deny a request—most of us make two mistakes that lead us to lose our temper and the other person's goodwill. The mistakes are:

- We think we can make the other person agree that his request should be denied.
- We focus only on this present confrontation with that person.

In this column, we'll see where these mistakes lead and how to correct them. Next time, I'll give you eight techniques for building a successful conversation on your new goals.

Where can you go wrong?

Mistake no. 1

When did someone ever change your mind about what you thought you deserved? If you think you can do this to someone else, you are almost sure to fail. Your frustration will lead you to lose your temper and, probably, insult the other person. The result: you will lose either him or his goodwill.

Mistake no. 2

Compounding your first error is mistake no. 2. If you see only the present confrontation and forget about your relationship with the other person, you are very likely to sabotage that relationship.

Later, when the immediate conflict has passed, you may find you have broken a valuable bond.

How do you say "no" successfully?

First, you recognize what you can and cannot do. You can choose your own attitude, tone, language, and goals. You cannot determine those of the other person. If his request means more to him than his job, his work with you, the prospects you offer him, and your belief in him—he will leave, no matter what you say or do. But—he too may have forgotten everything but the present conflict. If you remind him of the good things—your respect for him, his good prospects in the company, your honest regret that this request must be denied—you give him a real chance to accept the decision and go on working well.

Saying "no" successfully, then, means simply saying it with sincere respect and understanding.

To do this, you have to approach the conversation with two goals of equal importance:

- Short-term goal: To prevent the other person from doing something, while showing him untainted respect and goodwill.
- Long-term goal: To keep your relationship with the other person unharmed.

If you adopt these two goals, you will see several immediate changes.

Since you will have no unrealistic expectations, you will not be frustrated. And you will not lose your temper. Instead of focusing on changing the other person, you will be concentrating on your attitude, your words, your two goals. If you keep true to those goals, you'll feel good about yourself and the conversation. That positive feeling will have a better effect on the other person than would your negative feelings of frustration.

You will find yourself looking at the other person differently. Instead of facing a pigheaded egotist who is adding to your day's difficulties, you will see a person who feels deeply wronged. Since your goal is to show him respect and goodwill, you will think about his good sides. You will imagine how you would feel in his place. You will treat him the way you would like to be treated—with understanding, courtesy, and generosity of spirit. As a result, the person who came in feeling wronged will leave feeling that at least you understand him, feel for him, and consider him worthy of your time and consideration. He will feel better—even though you couldn't change the "no" to "yes."

Finally, since your goals imply caring for him, you may come up with a solution acceptable to both of you. As long as your only aim is to get him to see that you are right, you will be closed-minded. When you focus on those two goals, you will find your thinking changes. His needs will become as important as yours.

In January we'll look at eight techniques for conducting such a conversation with maximum success.

When you have to say "no" ...

Part 2

In my November column (p. **129**), I suggested that we make two major mistakes when we have to deny someone a request that means a lot to him. These are (a) believing that we can make the other person agree that his request should be denied and (b) focusing only on the present confrontation with him. Mistake No. 1 comes across as insensitivity. Number 2 appears as downright inhumanity.

If you determine not to make these mistakes, you will find the following 8 techniques for saying "no" easy, effective, and conducive to goodwill on both sides.

Remember your two important goals

I gave them to you in the last column. They are:

- **Short-term goal:** to prevent the other person from doing something, while showing him untainted respect and goodwill
- **Long-term goal:** to keep your relationship unharmed.

Keep both these goals firmly in mind as you talk, and you will establish the right attitude of *respect* and *goodwill.*

Don't expect miracles

If the request you're denying means a great deal to the other person, you may lose him no matter how you behave. But, you will not lose his belief and trust in you, personally. You may be glad of that in the future.

On the other hand, you may consider that losing that person is just not an option. Then, you may have to pay a price in order to keep him. You may be able to devise a compromise acceptable to both of you. Or you may even have to rethink the advisability of denying him this particular request.

If you don't expect miracles (e.g., that your conversational tactics will make your employee glad you promoted the other guy instead of him), you won't feel angry or frustrated when you don't manage to produce them.

Establish a positive ground

Begin the conversation by showing the other person your friendship, respect, and concern. Tell him how much you think of him. Refer specifically to the good things he has done. Of course, these remarks must be sincere. Saying "no" wouldn't be difficult if you hated the other person; there must be a lot you like in him. Don't be afraid to tell him. Your honest statement of friendship and goodwill will sustain him through the bad news and will help him realize that you are not denying him his request because of some personal animosity.

Give bad news clearly, with no hedging

Once you have established a positive ground, you can make it clear that your denial of the other person's request is in no way a rejection of *him.* Give him the facts clearly, completely, and with honest regrets that you have to disappoint him. Don't try to word the news in such a way that it sounds less noxious to him than it really is. Such pretenses will only enrage and disgust him further when he figures out the truth.

Explain your reasons

Even though the other person is probably reacting emotionally, his brain has not stopped working. If you explain why you must deny his request, his reasoning powers will take it in. He may not agree with your logic. But he will at least understand it.

And he will be a little grateful to you for having tried to explain.

Concentrate solely on your behavior

Although this is the most difficult technique, it is probably the most important. If you don't manage to do this one, you may throw away all the good of the other eight!

Remember, you have determined to adopt an attitude of respect, to show the other person goodwill, and to do what you can to preserve a good relationship with him despite the news you have to give him. He *cannot* stop you from doing these things, no matter what he says or does. Your attitude, words, and goals are entirely under your own control.

If his anger or resentment is bitter or sustained, however, it is easy to lose sight of your own goals and just respond with equal hostility. The result: Both of you lose. You doubly, because you give up both a valued relationship and your own stated goals for the conversation.

Resist any attempt to draw you onto the battlegrounds. Assert patiently that you understand his point of view and wish you had better news for him. Explain your reasons, more than once if necessary. *Always* show goodwill.

Remind yourself that you will have won if you have stuck to your own rules without getting dragged down into behavior you will later regret.

End on positive points

When the time for talking is up, stop politely. Reiterate your regret, goodwill, and desire to help the other person. If you can, offer some compensation. Show that you care; you may not think it matters, but it does.

Reward yourself

The confrontation was an ordeal, and you did the best you could. If you lost some control, at least you noticed it. You'll do better next time. Reward yourself for trying. You deserve it! □

Negotiation and communication

Part 1: Situation analysis

To negotiate means to try to settle differences. We have differences or conflicts all the time, but we don't always try to settle them. Somehow, it often seems easier to "live with" the situation, complain about it, sabotage the other side, or give up and withdraw. These are all *destructive* responses to conflict; the *constructive* response is negotiation.

How do we settle differences? By talking them away? Well, almost. The best negotiations (those that produce the greatest benefits for all parties) consist of rearranging or transforming differences until they become acceptable to everyone. This takes a great deal of careful communication, usually over a long period. So, in a way, good negotiators really do "talk away" differences.

Good communication, then, is one key to effective negotiation. The other is thorough, searching analysis and preparation.

Analyzing the conflict

The first step in any negotiation situation is to question your needs, interests, constraints, and major assumptions, particularly assumptions about the needs and interests of the other parties. Look beyond the obvious answers so you can devise creative solutions. The most important question to ask is *why*—not just once, but over and over.

Here's an everyday example. You are heading a lab that runs tests for other departments in your company's R&D center. Increasingly, you're being overwhelmed by "rush" projects. Apparently, 90% of the work the R&D center does is "urgent" and 10% is routine—or so they claim. Of course, nobody is happy with your lab because turnaround time on really urgent tests is unacceptable and anything submitted as "routine" is continuously pushed back until it's almost irrelevant. Rather than grumble, or quit and find another job before you develop ulcers, you decide to negotiate your way out of this demoralizing situation. Your immediate idea is to persuade the department heads to require a director's signature on any rush project. That should cut down on rush projects, you argue. Now, before you dash off to act on this, let's question your needs, constraints, and assumptions.

What's your need? "To cut the number of rush projects," you say. But *why*? "To improve turnaround for routine projects, yet get fast response to real rush jobs." Why do you want that? "To make the other departments more satisfied with our service." And why is that important? "It's our job: to serve the other departments." So, your real need is to serve your clients better.

Now your constraints. What's stopping you from bouncing projects back just as fast as they come in? "Staff size and equipment." The budget! If only you could have more people and more new shiny machines . . .

Finally, what do the department heads need? Reasonable turnaround on all projects so they can satisfy their clients. Fast response on urgent projects. Low cost. (Whatever money you get comes out of their budgets!) But they also have to keep their subordinates happy. One sure way to make them unhappy and unproductive is to give them more paperwork and less authority. And that spells big trouble for your "rush-buster" idea. Besides, suppose you did curb rush projects. Would you achieve your real goal, to give good, fast service to your clients? No. The real problem is slow general turnaround; that's why people submit everything as "urgent." And the only thing that would improve general turnaround is more resources for your lab. (Your studies show that productivity is not the problem, your lab excels in that respect.)

Putting it all together

What does all this mean? A new negotiation goal: Get a bigger budget. Who is the other party in that negotiation? Top management of the research center. What is the main obstacle to getting that bigger budget? The other guys, who are pushing their budgets as more important than yours. And what can you do about that? Negotiate!

Your analysis points you toward a two-step negotiation:

- Get the other department heads to support your request for more resources so you can give them better service.
- Get top R&D management to grant you those resources.

It won't be easy—a bigger budget is always a red flag. But if you succeed at the first step, your chances are hugely improved. We'll see next month how you might take that first hurdle.

Negotiation and communication

Part 2: Communicating differences away

Successful negotiation, I have suggested, hinges on *thorough preparation* and *good communication.* Last month, we looked at the initial preparation or "situation analysis" for a simple example: Your lab, which runs tests for other departments in the research and development center, is being overwhelmed by "rush" projects, resulting in low staff morale and strained relations with the departments you serve. Your first idea is to reduce rush requests by asking for a manager's signature on them. But as you analyze the situation, you realize that your purpose is more fundamental and requires a two-stage negotiation: Get the other department heads to support your request for more resources so you can give them better service; and get top R&D management to grant you those resources.

Let's examine how you might approach this situation and how good communication would help.

How to give and get cooperation

Negotiations reach the confrontation stage mostly because of *mistrust;* if you want cooperation, establishing trust is your most important job. And how do you build trust? By communicating consistently—not just at negotiation time when everything you say is suspect—that you are responsive to other people's needs.

In our example, to negotiate your colleagues' explicit support, you will need solid information on their needs and priorities, sustained efforts to meet those needs, and formal feedback from them on progress and remaining problems. (All this, of course, will form your backup documentation when you present your budget request to top management.) In other words, you are not dealing with some short-term negotiation; the approach required involves continuous high-quality communications, plus concrete actions, that establish your trustworthiness and willingness to cooperate.

Taking the cooperative approach seriously

Suppose you evaluate your current situation and find that:

- Your information on your customers' needs and priorities is sketchy and not based on explicit statements from them.
- You operate not on the basis of what would best serve your customers, but according to unwritten rules of "How a Good Analytical Testing Lab Is Run."
- You do not use regular feedback on how well you are serving your customers but only respond to complaints as they arise.

Clearly, this will not do. You decide to embark on a systematic communications program. First, you compose a memo asking for a meeting with your colleagues. The memo might start:

> I would like to ask your help in finding ways to improve the operation of the Analytical Lab. The staff of the lab is very anxious to do analytical work that pleases everybody, but we need better information than we have now. So, we would appreciate it if you could tell us about:
>
> 1. Any problems you and your staff have noticed
> 2. The things you need most from our lab, ranked by priority
> 3. Any extra services you would welcome.

At the meeting with your colleagues, make sure you keep quiet; you are there to listen. Your main contribution, besides the initial memo, is the agenda.

Next, you follow up with actions and information. Develop solutions to the problems that emerged and share those proposed solutions with your customers and get their responses. Keep trying to meet needs as they are formulated, get feedback on progress and remaining problems, and *document* each step.

Once you have shown your colleagues that you are committed to meeting their needs and you can document that the lab's productivity is tops, you have a good chance of getting their support as you approach top mangement.

Needs-centered communication benefits

All this may seem like a lot of effort, but it is worthwhile because it not only helps you reach your negotiation goal but also improves your work as well as your business relationships. In sum, good, regular communication will dilute potential conflicts, establish your credibility as a person who is truly interested in cooperating, and build better relationships.

Negotiation and communication

Part 3: How to conduct yourself in a negotiation session

So far, we have seen how to take the conflict out of a negotiation through careful communication and problem solving. In most cases you cannot achieve this in a day or even a week. You need systematic long-range planning. And you need actions, not just words, that demonstrate you are honestly trying to meet the other parties' interests and needs.

Gerard Nierenberg mentions an interesting example that illustrates this point.* In 1964, serious racial rioting broke out in New York; Rochester, N.Y.; Jersey City, N.J.; Paterson, N.Y.; and Elizabeth, N.J. But Newark, N.J., was spared, even though its large black population was coping with poverty and poor housing. In fact, when a proposed rally threatened to become a riot, the mayor was able to persuade 86 black leaders to help him turn the rally into a peaceful affair. Here's how he did it:

> The Mayor had made it a practice to keep in close contact with the leaders of the black community at all times. He made frequent visits to the Central Ward, the potential trouble spot. Through his consultations with the religious and civic black leaders, he was fully informed of the mood of the black community and well aware of black grievances.

If racial riots can be defused by the power of communication, how much more the everyday conflicts you face at work!

Showtime: five cardinal rules

No matter how well you prepare, negotiation sessions can be tense and full of surprises. But if you hold onto just five cardinal rules, you will achieve the best solution possible.

First, keep sight of your goals and priorities. Keeping a good relationship with the other people is always critical. In fact, it is usually much more important than "winning" the current round.

It is important to have a backup goal—what you will try to get out of the session if you cannot reach your primary goal. That way, you will avoid the feeling of total failure. For example, even if you cannot persuade your colleagues to go along with your solution, you may consider the meeting a success if you can get an agreement on what the problems are and a commitment to meet again soon.

Second, focus on problems, not feelings or personalities. Learn to spot the first signs of anger and resentment in yourself, then squash those emotions before they can take control. To steer everybody away from threatening emotions, periodically summarize the problem and review agreements you have all reached so far.

Third, listen carefully and calmly. No derogatory grunts as others speak; no raised eyebrows, smirks, or dismissive hand gestures. Instead, listen attentively to test your assumptions about the other parties' needs and interests. The better you understand the other side, the greater your chances of coming up with new proposals that might satisfy everyone.

Fourth, think before you speak. You do not have to answer every question on the spot. It you cannot respond convincingly, give yourself time; say something like, "That's a good point. I'll make a note of it; perhaps we can take it up next time." Or ask probing questions to amplify the subject. If someone makes an obnoxious remark, it is best to give no answer, but just change the subject.

Above all, do not argue or accuse; negotiation is persuasion—and arguing does not persuade. Using our example from last month, suppose a colleague snaps out, "I don't see why we should support your budget. We're not asking you to support ours." Do not immediately argue, "But you're not doing any services for my department, whereas my department is serving yours . . ." Rather, empathize, "I fully understand what you're saying. It is an unusual thing I'm asking. But clearly, we have a real problem getting the turnaround time you need, despite everything we've done so far—and I'm asking you to help me solve that problem."

Fifth, always see the best in people—and show it! This is the most important rule of all. Give them noble motives to live up to. Be generous with compliments. Empathize with their position. You don't persuade people by insulting them or battling with them. If you want to come to an amicable agreement with them, be amicable!

*The Art of Negotiating, Simon & Schuster, New York: 1968, p. 86.

Negotiation and communication

Part 4: Being a realist

When it comes to negotiation, humanity seems to fall into three groups: cynics, idealists, and realists. Cynics believe that everyone is out to bluff and cheat everyone else; the only way to survive is to bluff and cheat better than one's opponents. Idealists believe if one appeals to the good in people, they will always cooperate. Realists recognize that in most negotiations people's interests really are opposed, and that it takes hard work and creativity to reconcile those conflicting interests.

If you find negotiating disagreeable, imagine what the world would be like if we all had the same interests and had the same desires. When I do that, I can see negotiation as an interesting reflection of the variety of human life.

Know your game rules

Negotiation is one of the oldest social games; its basic rules have survived many civilizations. You have to go along with those rules to some extent. They are the "common language"—the frame people use to interpret what you say or do in a negotiation.

Here are some of the things people expect when they negotiate (however subtly) with you.

You'll start higher than you'll settle for

People just don't expect you to start with your lowest offer. So, whatever you propose at the beginning, they'll assume you will agree to less.

For example, suppose some clients ask you for a date of completion on a large project. If you start by offering your tightest deadline, the clients may press you for an even earlier date, trapping you in an impossible commitment. In the end, you may hurt not only yourself and your subordinates but also your clients, who count on you to meet the deadline.

And there is another point. Even if your first offer was a good deal, the people you're talking to may suspect that they got poor value just because there was no bargaining.

You will not be totally open about your needs, motives, and circumstances

There are several reasons "unconditional openness" is dangerous:

- It makes you vulnerable to exploitation by ruthless negotiators.
- Even if you are totally open, people will try to guess what you're hiding—so you really haven't gained anything.
- Openness can be an important bargaining tool—if you move toward it slowly and conditionally: "I'll give you some critical information; if you do the same, I'll give you more." When you start with total openness, you deprive the other side of an incentive to reciprocate.

Suppose you have a difficult proposal, such as asking your colleagues to support your request for a higher budget. Instead of opening with your proposal, wait until you're sure everyone understands the reasons behind it; then they may be more willing to consider your main request. And seasoned negotiators will feel more comfortable with you because you followed the conventions of negotiation.

You won't make concessions without being asked for them

If you do make an unrequested concession, people will assume that you believe your request isn't reasonable. They'll see it not as a concession but as a weakness.

Say you want an extra technician. Don't mention your "backup objective" of a "technician pool," or you'll get just that. Or, say you want more detailed information included in requests for analysis submitted to your lab. Don't immediately offer to try this new format for three months. Why go through the pain of "renegotiation" three months later unless you absolutely have to?

The realist view of power

Being a realist also means recognizing that no one but you will look out for your interests. But often we feel we cannot defend our own interests—we just do not seem to have the power to do it.

In fact, you have several major sources of power, including:

- Sheer persistence—holding on to your important goals and pursuing them flexibly and intelligently
- Appeal to indisputable moral or scientific norms
- The power to help people—to give them what they need
- Alliance with other people.

Draw on these four sources of power, and you will move closer to your goal. If your goal is good and reasonable and you are dealing with decent people, you will eventually achieve it.

Chapter 7
The Group Encounter

How to Run Successful Meetings and Give Winning Presentations

The columns in the first half of this chapter give you techniques for running meetings that accomplish goals and don't waste time. The columns in the second half show you how to give a fabulous presentation, *every time you get up to speak!*

How to stop hating meetings reveals three basic needs of all meeting participants and shows you how to fill them. A three-part series, **Making meetings work,** explains how to find and set the purpose of a meeting, write a complete agenda, keep the discussion moving, listen effectively, and conclude with real, measurable accomplishments. A two-part series, **Enemies of communication,** addresses two enemies of good meetings: **The hidden agenda** (Part 1), and **Baffle-gab** (Part 2). You'll discover how to recognize and avoid these foes. Finally, a five-part series, **Primer on meetings,** offers you techniques from the most successful meetings in the world.

The columns on presentation skills start with the number one problem: stage fright. The first two series of columns, **The secrets of public speaking** and **Enemies of communication,** give you practical techniques for stamping out that big fear and replacing it with a command of practical presentation skills. The next column, **How to give technical presentations without hating it**, concentrates on the *interesting* display of technical information.

And the **final column, A universal speech plan,** gives you a plan that will ensure the success of any presentation you ever make.

How to stop hating meetings

How do you run a meeting that really works? A meeting people-including you—won't hate? The answer is simple.

Recognize the needs of the participants and fill them.

Here are three key needs of participants and techniques you can adopt to fill those needs.

Participants need to know, explore, and accomplish the meeting's stated objectives

The main cause of fuzzy meetings is lack of sufficient preparation. You can focus your participants on the important points and gear them to prepare themselves properly by **circulating an agenda 48 hours in advance.** Make copies to pass out at the meeting, too.

Divide the agenda into two sections: **Action Items** and **Discussion Items.** Be sure to complete all Action Items before you move on to the Discussion section.

Next to each item, put the **name of the person who will introduce it.** Before you circulate the agenda, remind each person listed to prepare the following:

- **A summary statement** of the topic
- **Essential** background information (facts people need to understand the topic and its significance)
- **A suggestion** on what to do about the item
- **Nonessential but useful** information (facts the speaker needs to answer questions people might have).

By putting a name to each item, you give the person the responsibility to prepare his or her part. And by circulating the agenda with the names on it, you let everyone else know who is responsible for what.

Schedule realistic **times** for each item. Try to follow your schedule, but note **in advance** which items you can afford to omit or cut short if you run out of time.

At the meeting, stick to the agenda and sequence.

Ask people to raise a hand to get your attention when they want to speak. In this way, you control the discussion and can turn from meeting-hoggers to the shyer members whose ideas should also be heard.

After discussion on an item is complete, **sum up** the conclusion to be sure all agree. **Never** omit this step. It clarifies exactly what you have decided and lets participants know **what was accomplished.**

Participants need to meet their personal objectives

Have you ever noticed how the second part of a long meeting or workshop seems to go more smoothly than the first? Do you know the reason? The coffee break! During that time of casual talk, people often feel free to discuss their own real concerns—such as protecting their department's interests or fearing someone else will steal credit for their ideas.

You can give participants the chance to seek their own objectives by making your first item on the agenda an **open discussion.** Tell people they have five or ten minutes to talk freely on any item on the agenda. To avoid a free-for-all, establish the rules you would have for a brainstorming session. Anyone may say anything about the subject, but there is to be no criticism or personal attack. People may talk around the table or in small groups.

When the discussion period is over, call the meeting back to order. Ask for comments or suggestions that grew out of the discussion. Note these and address them when the subject comes up on the agenda.

This short open discussion can be repeated any time you find the meeting getting out of control. Remember, the personal motives won't go away. Recognize them, give them some rein—and you'll be amazed at the cooperation and enthusiasm that result.

Participants need to keep the meeting as short and pleasant as possible

How do you deal with the people who disrupt meetings with lengthy monologues or interruptions? Silencing them rarely works. They either ignore your call for order or subside into a steady, disturbing buzz.

But **listening to them** almost always works. The disrupter wants his or her words to be **heeded, understood,** and **seriously considered.** Fill his needs—and he is very likely to fill yours and stop.

Listen carefully to the disrupter, with an intent **only** to understand. Say you want to recap what he has said to be sure you have understood. Once he's satisfied that you've got the message, do one of the following:

1. Address it directly, saying what you will do.
2. Ask for a **timed** discussion, maximum five minutes.
3. Tell the speaker you need more time to consider this and will take it up with him at the break.

If you can meet these needs, no one will hate your meetings. In fact, everyone will look forward to them. Even you!□

Making meetings work

Part 1: The purpose of a meeting

Meetings. Often their most striking product is a corporate groan. People resent them. They take so much time and energy to yield such meager results. They're also expensive. Consider a three-hour meeting of ten executives, for whose time the firm pays each $100 an hour in compensation and benefits. That meeting costs *$3,000*. Yet we continue to hold them, day after day, week after week. Why?

Because we need to work together. It's that simple. People working together for a common goal have to come together to discuss their efforts to meet that goal and the problems that get in the way. Meetings can dramatically raise group morale *and* productivity. The trouble is — they usually don't.

In this three-part series, we'll see how to *make meetings work*. First, we have to consider why we have meetings. What is their purpose? Next, we'll examine the tools of communication that enable the user to fulfill that purpose. And finally, we'll see how to solve the biggest problem of meetings: the feeling that "Nothing was accomplished."

Reasons vs. Purpose

Many people confuse the *reason* for calling a meeting with the *purpose* that meeting serves. Let's look at some reasons first.

We often have meetings to SOLVE PROBLEMS. In fact, this is the most common reason for calling a meeting. It's even reflected in the latest Business Buzz Word used in some companies today. In these firms, you don't have meetings any more; you "take" them. It's like taking a pill. If you have a problem you can't overcome alone — try taking a meeting. It may indeed cure your ills!

Another reason we have meetings is to REVIEW PROGRESS ON A PROJECT.

A third is to BRAINSTORM — to try to put a lot of good minds to work together to come up with something new.

A fourth is to GET TOGETHER TO SHARE IDEAS, PROBLEMS, SOLUTIONS, AND GRIEVANCES. This is one big reason for the existence of conferences.

And finally, we may call a meeting simply to GIVE PEOPLE INFORMATION.

These are some of the reasons we have meetings. But the meetings themselves will work only if we keep in mind the *purpose* a meeting serves.

> The purpose of any meeting is for people to accomplish something together that they could not do so well separately.

When the leader of the meeting forgets that purpose — you can forget about the meeting. It won't work.

A simple meeting that should not fail . . .

Let's take an example. The simplest type of meeting is the one you have just to give people information. You can often do this best in a meeting because:

- You save time — you only have to say it once to everyone.
- You can be sure all the people hear the information as you want them to hear it — not adulterated by additions and subtractions along the grapevine.
- You have an opportunity to see people's reactions and respond to their questions — and to avoid having to answer the same question on many occasions.

It seems as though this meeting could not fail. You are just going to assemble people to give them some information and answers.

But *it can fail* — if you forget that the purpose of any meeting is for people to accomplish something together that they could not do so well separately.

Forgetting the purpose = failure

If you forget this purpose, you may decide, for instance, that you will hold this meeting purely to transmit your information as quickly as possible. You may then proceed to call several people away from their own jobs, sit them down, deliver your message, and dismiss them. You leave no opportunity for questions or discussion because your purpose is to *get your information across quickly* — and question periods are notorious time-stealers.

This meeting will almost certainly bomb, for several reasons.

First, when responsible people are called away from their jobs, they naturally expect the demand for their presence and time to be justified. *Saving you time* is not a sufficient justification. In fact, it shows lack of respect for them and their work. They will resent you and probably receive your information with bad grace and scant attention.

Second, if you don't allow time for questions, that doesn't mean the questions don't exist. People may ask for clarification individually, taking more time than a question period would require. Or, miffed at your summary announcement, they may leave you alone and decide by themselves what you meant and what they're going to do about it. The result will be much error and little goodwill.

Finally, by omitting feedback, you deny yourself the chance to learn how your information affects the people and the work of your company.

You don't always need a meeting

But — suppose you have some information to give that people cannot change or adapt and are fairly sure to understand. And suppose *your* purpose is to transmit this information to them as quickly and efficiently as possible. DON'T CALL A MEETING. Write them a memo. It won't take you any longer — and you won't have a failed meeting by thwarting the inherent purpose of the meeting.

Meetings have a single purpose: to enable participants to accomplish something together that they could not do so well separately. Next time, we'll look at some communication tools that will help you fulfill that purpose at every meeting you hold.□

Making meetings work

Part 2: Communication tools

The communication tools that you use to make your meetings work all function to serve the purpose of a meeting. That purpose, as I stated in the last column, is simply: *to enable participants to accomplish something together that they could not do so well separately*. They can do this in a meeting by:

- Sharing a large task among themselves
- Offering many different solutions
- Picking up each other's ideas and turning them into new ones
- Bringing several specialized types of knowledge to one problem
- Airing all reactions and questions at once, in one place.

These are only a few of the ways people use meetings successfully, but they alone show you the kinds of communication tools you need. You want to have *all* the participants actually participate, and you want to reach a common understanding of certain facts or decisions by the end of the meeting. Here are five tools to help you.

1. Write a complete agenda. List the topics you want to cover during the meeting, in the order of their importance. Then look at your list and decide whether you really can discuss all that in the time you have allotted. Be realistic. One of the greatest causes of frustration in meetings is the long agenda that cannot be adequately covered in the time allowed.

Once you have pared your list down to a reasonable four or five items, see if you can be more specific about each. Each item on the agenda should be written out as a *directive*: a statement of what you want to do about that topic at the meeting. If you've written:

Late delivery of parts by Supplier X

you can be sure to waste many minutes of meeting time listening to various participants grumble about how they personally suffered from this consistent problem. Instead, write:

Find ways to speed up delivery of parts to Supplier X.

The people's minds will be focused on action, not complaints.

Once you have written your agenda, note *for yourself* the amount of time to be spent on each topic. Don't write the suggested times on your participants' agendas; you'll make them feel they're back in grammar school, and you'll stifle the less garrulous ones from contributing. But *you* should know when to bring discussion on one topic to a close.

Finally, with the times in front of you, find the midpoint of the meeting and insert "Mid-Meeting Review" at that place on the agenda.

If possible, distribute this agenda to your participants some time before the meeting. Ask them to prepare to contribute their ideas and experience on each topic.

2. Consider the personality and interests of each participant. Write down the participants' names and put a star by the quieter people who will be attending. Plan to keep this list with you at the meeting and make a point of asking those people what they think about the topic under discussion. They will give you a better, more confident response if you have prepared them to think about the topic by giving them your agenda well in advance of the meeting.

Now note the interests or areas of expertise of the participants. This small act of preparation will help you call on the appropriate person for ideas or information on a particular subject. For instance, someone very involved with organizing community or political projects might be just the one to come up with a creative idea for speeding up those deliveries. The problem is not unlike those he or she has had to solve in these extramural activities. But neither you nor that person may make the connection unless you point it out.

Finally, you can use this short "personality analysis" to turn parts of the meeting over to those whose specialty the topic is or to those who are more intimately involved with a particular subject. This device will add interest to your meeting and keep your participants alive!

3. Pass the ball. When one person becomes overly talkative and threatens to hog the meeting or keep it from moving on, you can start talk circulating again by "passing the ball." This means simply that, when the talker pauses to breathe or to marshal his thoughts, you say:

So you think we should just leave Supplier X, Joe? That's an interesting possibility. How do you feel about it, Brad?

You encapsulate the speaker's remarks into a general statement—and pass it on to another participant for comments. The garrulous member cannot be insulted, for you have obviously listened to and understood him. And the meeting gets moving again.

4. Treat each speaker with friendly respect. This is the biggest and best cure for the deadly forming of factions in your meeting. There are almost always at least two sides in a meeting: the person chairing the meeting vs. everyone else! When you assemble people, they tend to be on their guard, wondering if what you have to discuss is worth the precious time they have given up and — more important — wondering what you will expect of *them*. If you respond to each comment, no matter what it is, with attention and friendly respect, that speaker will feel you are on his side. Even if you hold opposing views, he will feel empathy with you, and the barriers will start to crumble. It never fails.

5. Listen to understand. When someone else is talking, listen *only* to understand. Not to nitpick, catch out, criticize, or point out inconsistencies. You can check your listening skills by determining always to repeat the essence of each speaker's message for the group. In this way, you will make sure that each participant's contributions become an active part of the meeting.

Use these five simple communication tools — and you'll find your meetings changing from aimless carping sessions to real group accomplishments.□

Making meetings work

Part 3: Solving the biggest problem

The biggest problem of all meetings — in business, community, or politics; big or small; formal or informal — is the frustrated feeling at the end of the meeting that "nothing was accomplished." Human beings seem to have an endless capacity for sitting around together, getting involved with something, talking excitedly about it, and resolving — nothing! Then they have to schedule another meeting to try to figure out together why this one didn't work. And so on, *ad infinitum.*

Your meetings can work. Every meeting you hold can accomplish a great deal, if you will just follow four rules. Here they are.

(1) Circulate an agenda of specific, not general, topics, and consider each separately. Suppose just for a moment that your department makes patchwork shirts — each one different — out of recycled materials. Demand for the shirts has increased, and you're having trouble getting a regular order out to a steady client on the dates he has specified. You call a meeting to resolve this problem.

Your agenda should NOT look like this (though most do):

> To correct consistent late delivery of shirts to Company X
> 1. Present problem.
> 2. Discuss with all present.
> 3. List possible solutions for future implementation.

With number 2, you'll engender a free-for-all or a dead silence. Either way, you'll never reach number 3!

To make your meeting work, you must spend some time before it thinking carefully about the problem. Break it up into manageable segments. Then you can make your agenda a list of these segments.

For this meeting, you might consider: Who in your department is involved in any way in the expedition of this order? What outside departments or firms contribute to it? What forces could be causing the delay — e.g., complacency with product's popularity, eagerness to build up new business at the cost of maintaining steady customers? Where in the process of development of the product are there potential delays? Now you can turn these questions into an agenda:

> 1. Note *schedule, time required,* and *any regular difficulties* occurring in the receiving of scraps (Mary), laying them out for selection (John), choosing scraps for each shirt (Bill, Diane), making shirts (Amy, Bob, Kurt, Diane), and preparing finished shirts for delivery (Mary, John).
> 2. Delegate people to check with Departments A, B, and C and Supplier Y to find out schedules, actual dates and times required, and areas of possible or consistent delay.
> 3. Discuss other forces that could lead to delay (10 minutes max.)
> 4. MID-MEETING REVIEW
> 5. Note possible alternative departmental scheduling for each phase of production. (See item 1 on agenda.)
> 6. Note possible alternative delivery arrangements and sources of raw materials.
> 7. Consider adding people to any area where delays always occur.
> 8. LIST DISCOVERIES, SUGGESTIONS, and DECISIONS MADE AT MEETING.

Note that the first part of the agenda consists of *fact-finding.* You do a mid-meeting review to draw the facts together into a meaningful constellation. The second part works with the facts to *change* present circumstances.

(2) Write down each decision or suggestion as it occurs and list them all at the end of the meeting. Very often the feeling that "nothing was accomplished" at a meeting is not based on facts. Suggestions *were* proffered, decisions *were* made — but they got lost in the hubbub. Don't leave the note-taking to your secretary. Keep your own ears open for suggestions and decisions and note them on paper as they are made. Then at the end of the meeting, you can tell the participants exactly what the meeting has produced.

(3) Focus on the topics to be discussed, not their background. Sometimes more than half a meeting is lost to long-winded dissertations on the history of a problem, the making of a product, or the long-standing inefficiencies of the people involved. These remarks serve your purpose only when they contribute to the resolution of the problem that brought you all together. If they don't, you have to cut them short.

For instance, in this example, someone may be very eager to expound on the elaborate methods of recycling soiled or spoiled material. He may do this either because the subject fascinates him or because he wants to draw the focus away from his own contributions to the delays! Whatever his reasons, YOU must focus on the topic. Don't silence him thoughtlessly. He may be telling you that the time required for recycling is longer than you had imagined or that there is a quicker method used by an alternative supplier. Don't be afraid to interrupt to bring him back to your topic, and keep noting the precise facts or suggested actions you draw from his talk.

(4) Hold all discussion to the topic at hand. This means curtailing the chair-hoggers and drawing out the silent experts . . . interrupting personal discussions . . . cooling animosities. You'll find techniques for doing all this in Part II of this series, "Communication Tools" *Tappi Journal,* Sept. 1984, p. 142. If you strictly stay on each topic, interrupt politely but firmly when speakers drift off, write down all discoveries, suggestions, and decisions, and present what has been decided and what is left to decide at the mid-meeting review — you'll be a magnetic force that holds the meeting on the line of the agenda. And when you read out the list of discoveries, suggestions, and decisions at the end of the meeting, you'll know that you've solved "the biggest problem," forever.□

Enemies of communication

Part 1: The hidden agenda

"Excuse me — I just heard what you're thinking, and I don't agree."

It's too bad we don't say that more often. Absurd as it may sound, it is exactly what we often feel during a conversation. And it's the reason for so many failures at communication in business.

We react to what *we* think the other person is thinking.

Let's take an example.

Jim and Brad are working on setting up a training program for their company. Jim believes the program should stress practical, on-the-job experience. Brad contends that solid, formal training sessions should precede any company work. At the moment, they're planning a lecture on financial basics for new employees — how to write budgets, organize expense accounts, and so on. They have met to decide whether to use a member of the company's accounting department to give the presentation or to bring in someone from an outside training firm.

Jim says, "I don't know that a member of an outside firm will understand our particular budgeting methods and needs as well as someone who's regularly working on our books."

If anyone else had made this point, Brad would probably have considered it thoughtfully. But, as it comes from Jim, Brad immediately thinks he hears a sly thought behind it. The thought goes: "Aha! This is the way to get the trainees out of the lecture hall and into the accounting offices. If we give the job to one of our accountants, I can encourage him to say just a few words to them and then give them some practical work to do!"

Reacting to what he thinks Jim is thinking, Brad rejoins, "Oh, I think a member of a training firm would be much better. Those people don't just know how to do accounting — they know how to teach it."

And, of course, Jim hears: "Aha! This is my chance. If I get a professional trainer in to give this lecture, I'll have taken the first step to making the whole program nothing but a series of formal lectures." So, Jim polarizes his position. He states categorically that no member of any training firm ever had any practical accounting experience in his life. Brad counters that this is nonsense and that, anyway, there would be less mistakes in accounting if the staff had a more solid formal education in the fundamentals of finance. The upcoming presentation gets lost in the foray. And nothing is accomplished.

This scenario is typical of the many useless meetings that turn into frustrating funnels of everyone's time and energy. How can you avoid them?

Three steps for coping with hidden agendas

1. *Check yourself before the meeting.*
Imagine the participants in the meeting and tell yourself what you are afraid each one will try to accomplish. Remember — the participants include yōū! What's on your hidden agenda? Be aware that you will tend to see each person's particular goal behind *everything* that person says. And that the others will do the same. Building this awareness is the most important act in bolstering real communication.

2. *If you are running the meeting — or if there are only two of you — begin the meeting by sharing this awareness.*

Tell the others what you believe they want and what you want. This simple statement will bring the hidden agendas out of their closets and render them less powerful! Others may refute your claims; accept their statements. Then suggest that, whatever each one wants, you all try to forget these personal goals just for this meeting. That you all try to concentrate on the subject to be discussed and its own inherent merits or flaws. And that, if anyone seems to be using the discussion to further his or her own goal, the rest of you *ignore* this phenomenon and concentrate only on the actual statement he or she is making.

3. *Do everything you suggested in 2.*

If there was no opportunity to share your awareness with the others — take your advice yourself anyway. You'll stop the polarization game, because polarization needs two poles. You won't be reacting to what you think the other person is thinking or planning. By keeping your focus on the subject under discussion, you will act as a sort of gravitational pull for the other speakers who may be starting to spin off in unproductive directions.

A road to new solutions

Is there a niggling doubt in your mind as you read this? Are you wondering about the consequences — how such behavior will affect the goals of *your* hidden agenda? Are you afraid that if only *you* determine to focus on the subject, the others will stealthily arrange for their unstated needs to be met?

Your doubts are reasonable but answerable. First, if you keep your attention on the subject, not on the many hidden agendas, you will see much more clearly what is happening to that subject. In fact, you will have more influence on its fate, because your attention won't be diverted. If a speaker is using the subject to meet other needs, you can always step in and question his motives or the value of his suggestions. But only insofar as that motive or those suggestions affect the subject or problem under discussion.

Suppose you all conclude that the best resolution of the problem is one that could, eventually, fulfill someone else's hidden agenda — not yours. Then be honest. Say that your agree with this particular solution but that you are worried about its implications. You can be sure that your honesty and willingness to cooperate will make other people far more willing to meet you halfway, on this or other more important issues.

And there's one more point. By temporarily ignoring the hidden agendas (your own included), you may see the subject in a different light. Brad may have decided that Jim was right: the trainees could learn more relevant information from an in-house accountant. Accepting this, he could then have gone on to consider the possible choices and the format of the session. He may even have concluded that a compromise between theoretical lectures and some practical or simulated experience would be the best way to present the financial picture.

We all tend to go into meetings or discussions convinced that, on one point at least, we alone are right. If just one person lets go of this conviction for just one meeting — it's suprising how many new solutions appear.□

Enemies of communication

Part 2: Baffle-gab

'Twas brillig, and the slithy toves
Did gyre and gimble in the wabe;
All mimsy were the borogroves,
And the mome raths outgrabe.

—Lewis Carroll, *Through the Looking Glass,* Chapter 1, "Jabberwocky."

The most striking effect of Lewis Carroll's nonsense lines is: they seem to make sense! You *almost* see something — a primordial landscape, perhaps, or maybe the sea and the earth moving restlessly before some strange, imminent calamity.

But actually, the words you think you somehow comprehend are slippery beings. Get too close, and they vanish or turn into something else. Does "brillig" suggest *brilliant* . . . or *thrilling*? Are "slithy toves" reminiscent of *slithery doves* . . . or *slippery coves* . . . or *slimy tomes*?? By the time you try to make each nonsense word mean something, you'll have a headache and you'll realize that you can't demand clarity on the other side of the looking glass!

Unfortunately, too many people behave as though they had entered that fabuland as soon as they step into a business meeting. Suddenly, their English words turn into baffle-gab. They jabber. And their listeners feel that something must have been said, because the words seem to resound with meaning. But if anyone stops to examine the multisyllabic output — too often he finds himself slipping and sliding among mome raths and borogroves! Because the meaning that seemed to be there — isn't.

The sounds of business

Here are two declarations overheard at corporate business meetings.

Speaker A. Not having been involved in the real interactive management of the workings of this thing, I think I can offer a personally disinterested viewpoint on the matter.

Speaker B. While your criticisms of the nature of the Committee's deliberations may be to a certain extent justified, I feel impelled to state categorically that our final decisions were the result of a panoply of data drawn from a vast network of regional resources, all of which was channeled and integrated into a formal presentation that was made available for the study and eventual objective judgment by all Committee members.

Let's see what is common to these typical outpourings of baffle-gab. First, they use expressions that appear to make sense but that, on examination, simply dissolve into twaddle. What kind of management is "interactive" and what is not? What does "the workings of" add to "this thing"? If a person's viewpoint is "personally disinterested," *whose* point of view is he taking? And by "disinterested," does he mean *uninterested* (Webster's first meaning of the word) or *unbiased* (meaning number 2 in Webster)?

The list of demands for clarity is even longer for Speaker B, who got so mesmerized by the sound of his own words that he finally ignored their meaning altogether. *Panoply* has as its first meaning, "a full suit of armor" and as its last, "a display of all appropriate appurtenances." But the speaker didn't mean either. He probably *meant* "all the available data" or "a considerable amount of data." But the word "panoply" *sounded* so full and awesome, he couldn't resist it. It had a Greek, learned character. It called to mind the wideness of *panorama* and perhaps the sense of hugeness and power captured by *monopoly*. And — the speaker probably figured his listeners understood the word no better than he did!

What all this means is that people who use baffle-gab are not really concerned with what they're saying. Their main aim is to produce impressive sounds. But for the conventions of business, they would do far better to enter the meeting with a trumpet fanfare. That way, at least, there would be no verbal message to be hidden in their sonic exuberances.

Surely we should be doing more than just sounding off. Here are some ways to bring communication back to your business, and get your people back on the right side of the looking glass.

How to replace baffle-gab with meaningful talk

To further communication and eliminate baffle-gab, you can—

Recognize it in yourself. You'll know it before you speak, for it gives ample warning. You'll feel suddenly that you *must be heard* and, furthermore, that you must be *taken seriously* and *given great respect*. At the same time, you'll be vaguely aware that you haven't a lot to say or that your point is not as strong as you would wish it to be. This combination — the desire to be heeded and the lack of a power-packed message — almost invariably leads to an attack of baffle-gab. Resist it. Don't say *anything* until you feel comfortable just making your small point clearly. You may be surprised at its effect. A single white feather positively gleams on a heap of garbage!

Recognize it in others. When someone uses a lot of words that don't yield any clear meaning — *tell him you don't understand.* You don't have to be unpleasant about it. Just tell him honestly. Addressing Speaker B, you could ask him what type of data has been collected . . . what regional resources were used . . . and what ensured that the Committee's judgment was indeed objective. One person's desire for truth can puncture all the baffle-gab filling up the boardroom air.

Discuss these points with your staff and colleagues at a meeting. By making everyone aware of the prevalence and uselessness of baffle-gab, you may launch a departmental campaign to end it. Your meetings will be far more meaningful.

If you don't get rid of the baffle-gab, each meeting may quickly become just

a tale . . .
Full of sound and fury,
Signifying nothing.

Primer on meetings

Part 1: A primary purpose that inspires and unites

One type of meeting that has been phenomenally successful all over the world is the AA (Alcoholics Anonymous) meeting. Most alcoholics cannot beat alcohol by themselves or even with the help of doctors, psychologists, loving relatives, and ministers — but **solely by attending AA meetings**, they recover from alcoholism and are restored to a healthy, productive life. To anybody who regularly attends business meetings, this has to be a miracle.

I recently had a long talk with a friend who is in the AA program. I wanted to know what makes these meetings so powerful. What I learned made me revise my thinking on how we should run meetings in business.

Two ingredients of AA meetings impressed me especially:

1. They all have the same powerful, uniting **primary purpose**.
2. They have a **strong, simple structure,** based on unwritten rules on how to behave in an AA meeting.

Both these ingredients hold great lessons for business, I believe. Today, I want to share with you my thoughts on the primary purpose. Next time, we'll see what we can learn from the structure of an AA meeting.

An unswerving focus on the primary purpose

Most AA meetings have a specific **topic**, such as "anger," "fear," "keeping it in the now," or one of the 12 steps of the program. Now, if they were business meetings, one might demand that they "define the problem," "generate a plan," or "produce an agreement." But that's not what makes an AA meeting good. Rather, a good AA meeting achieves the primary purpose of AA: to keep the members sober. Every meeting has this same primary purpose.

"Doesn't this get boring and repetitive?" I asked my friend. "You've been happily sober many years. Do you still go to meetings to 'stay sober another day'?"

"Not quite. But the meetings do move me further away from a drink. There's no way to stand still in **AA.** You either move away from a drink, or you move closer to picking up a drink. You have to **keep growing** as a person, or the same character defects that caused you to drink in the first place will make you drink again."

Now, what really astonished me was that people apparently never lose sight of this primary purpose of a meeting. And why is that? **Because they read a Preamble that reminds them of the primary purpose.**

A primary purpose for every business meeting

The more I thought about these things, the more I could see how well they really apply to business. Consider some of the similarities:

- Every business also has a primary purpose that ought to be supported by every meeting of its people, no matter what the specific objectives of the meeting.
- Like the alcoholic, every organization that isn't constantly improved will get worse and eventually perish. It will **not** just stay the same. (This is the central idea of kaizen and Total Quality Control.)
- The primary purpose of the organization must be **inspiring** and **constantly and powerfully communicated.**

Now, giving every new employee a booklet on "What XYX Corp. Stands For" is not a powerful way to articulate your primary purpose. What's the alternative? **The Preamble.** More specifically, a Preamble like this:

> *The primary purpose of this meeting is to improve the way we do things here so that we, and the people with whom we deal, will be more satisfied. Our guiding principles for every meeting are honesty, respect, equality, and openmindedness.*

The ritual power of the Preamble

I'll have more to say about this Preamble in the following columns. For now, just imagine what would happen if every meeting in your group or organization started out with this Preamble, followed by a statement like: "Our specific objectives today are..." Don't you think the tone and focus of your meetings would soon start to improve dramatically?

Such preambles are powerful rituals. They clear people's minds of clutter and signal the start of serious business. Many legendary figures of business started their meetings with a prayer (a religious form of Preamble); they felt that this contributed greatly to the success of the business.

I experienced the power of prayer at a school for juvenile delinquents where I taught. All the teachers had great trouble getting control of the girls in our charge—except the nuns. And **what was their secret? You guessed it: They started every class with a short prayer.**□

Primer on meetings

Part 2: The structure of a successful meeting

Last month, I suggested that we can learn many valuable lessons from one astonishingly successful type of meeting — the AA (Alcoholics Anonymous) meeting. Two reasons these meetings work so well attracted my attention: (1) they never lose sight of their Primary Purpose (to keep the members sober); and (2) they have a strong structure based on a few unwritten rules and traditions.

Last time we discussed a simple tool (borrowed from AA) for reminding business-meeting participants of the Primary Purpose of the meeting. That tool is a Preamble such as:

> *"The primary purpose of this meeting is to improve the way we do things here so that we, and the people with whom we deal, will be more satisfied. Our guiding principles for every meeting are honesty, respect, equality, and open-mindedness."*

This time, we'll look at the structure of an AA meeting to see which elements we could borrow to improve our business meetings.

The structure of success

Just what happens at an AA discussion meeting that makes it so effective? Here are some elements that intrigued me especially:

- Leadership of every meeting is shared between a chairperson (elected for a short cycle of 3 to 6 months) and a leader for just that meeting.
- The chairperson reads the Preamble, introduces the leader, collects money, and takes care of other "housekeeping." The leader (who may be only 3 months sober) briefly shares on an AA-related topic, such as "gratitude," "fear," or one of the 12 steps of AA.
- Other members then take turns talking about the topic. The leader may reply briefly to contributions, but the priority is on letting everybody share. If time runs out, the leader may ask for a show of hands so everybody who needs to talk gets a chance to do so.
- There are certain taboos. Above all, don't criticize people (except yourself) directly; don't preach or tell people what they "should" do; and generally don't play "Big Shot."
- There is an atmosphere of refreshing honesty and trust that allows people to share extremely personal things, either to "unload" or to help others who face similar problems.
- At the end, people thank the leader for a "good meeting"—especially if the leader is a newer member.

All this has a combined effect of keeping the members sober until the next meeting. Why? Because they're able to:

- Renew hope that they're getting better in every respect
- Realign their thinking to remember what's important
- Get honest with themselves as they take an inventory of their attitudes and character defects
- Feel supported and accepted by the group
- Build their self-esteem by helping others.

Just compare this to what goes on in the typical business meeting: sniping, power plays, blaming, fudging, pulling rank, showing off, hogging time, pulling down people's ideas. After that kind of exercise, no wonder we feel battered, hopeless, and resentful.

A five-point program for a successful meeting

Wouldn't we all rather feel hopeful, renewed, and appreciated after a meeting? Then, why not adopt some of AA's practices? They make eminent sense in ANY setting. Here is a five-point program to set direction (I'll have more to say about it in the next columns):

1. Share and rotate power, especially for your routine meetings — and always have both a chairperson or facilitator as well as a meeting leader.
2. Always start by reading the Preamble, to remind everyone of the larger Primary Purpose (as opposed to the specific objective) of the meeting.
3. Make sure everybody gets a chance to contribute.
4. Avoid all direct criticism (except "self-inventory"), preaching, and displays of "authority."
5. Support other members of the group—especially newer ones — and generally keep a positive spirit so everybody leaves the meeting feeling better, not worse, than before.

What does it take to put this into practice? Commitment, first of all. Courage to set an example. And probably some training to start turning things around. But with so much at stake, it's well worth a big effort from you and everybody in your group.☐

Primer on meetings

Part 3: How to run a good meeting

In the last two columns, I discussed things businesses could learn from one of the most effective meetings around: the Alcoholics Anonymous (AA) meeting. We saw how a meeting **Preamble**, modeled after AA's Preamble, could help focus everybody on the **primary purpose** of the group. We looked at some unwritten rules of AA meetings that would improve our business meetings:

1. Share leadership responsibility with chairperson and leader.
2. Let everybody have a chance to contribute.
3. Avoid criticism and preaching.
4. Be honest, trusting, and caring.

This time, I'd like to share thoughts on how to chair or lead a meeting. As in AA, these two should be kept separate. The chairperson picks a speaker and ensures the meeting follows the traditions established; the leader moves the meeting along and makes contributions.

The many purposes of a meeting

A good meeting is a meeting that achieves its purpose. And what's the purpose of a meeting? To solve whatever problem has found its way onto the agenda? Yes—but what if in the process of solving that problem, we damage our relationships with others or demoralize the group? Then the meeting has missed its real purpose.

A meeting has as many purposes as you care to give it. If you want to make meetings a healthy part of your group's work, consider the following purposes:

1. Bring the group closer together.
2. Air problems that affect the work or the group.
3. Build people by developing their self-esteem and skills.
4. Refocus everybody on the principles of the group.
5. Achieve the specific task at hand.

Running a good meeting means keeping **all** those purposes in mind. Let's see now how the chairperson and the leader can contribute to achieving these five purposes.

Purpose 1: bring the group close together

- Chairperson: read the preamble; set the right tone when introducing the speaker; call attention to the group's "meeting traditions" if things get out of hand.
- Leader: let everybody contribute; treat all participants as equals; always remember that the leader is a "trusted servant of the group," not a power figure.

Purpose 2: air problems that affect the work or the group

- Chairperson: read the preamble: encourage the leader to deviate from the agenda if serious problems come up that demand immediate discussion.
- Leader: set an example in his or her own contributions; avoid criticism of ideas; gently but promptly cut off judgemental responses to a speaker's comments.

Purpose 3: build people by developing their self-esteem and skills

- Chairperson: select a leader for the meeting who would benefit from the experience of leading.
- Leader: encourage shy participants to contribute; avoid criticism and preaching; be supportive and appreciative.

Purpose 4: refocus everybody on the important principles of the group

- Chairperson: read the preamble; set the right tone when closing the meeting.
- Leader: tie the specific objective of the meeting to the primary purpose and the general principles of the group; restate or summarize people's contributions in a way that clarifies their connection to the bigger principles.

Purpose 5: achieve the specific task at hand

- Chairperson: select a good leader for the meeting; help with selecting other meeting participants, if necessary.
- Leader: prepare a good agenda; get the right people to attend the meeting; help key questions and periodic summaries of contributions; solicit and summarize action steps; assign clear responsibilities.

You may object to "Achieve specific task at hand" as the very last purpose. We're not in business just to make each other feel better. We need to "get outstanding results" and "solve tough problems." I agree. But we get outstanding results **only by functioning well as a group.** By paying more attention in our meetings to the way we interact as people, we'll come away inspired, renewed, and hopeful—**and** we'll solve all those problems better than we do when we focus on them alone.□

Primer on meetings

Part 4: Building a culture that fosters good meetings

Probably the biggest factor determining the quality of your meetings is the culture of the group. Culture sets expectations—and as smart teachers and managers know, people tend to live up to expectations.

Changing culture—the quiet way

But how do you build a culture that produces the right expectations in people? This may seem a huge task calling for extreme measures—reorganization, speeches and presentations to the troops, task forces and committees, new personnel policies, surveys and studies by outside consultants . . . Unfortunately, such massive efforts rarely wipe out the prime enemy of change: cynicism. In fact, the harder you "sell" your message of "cultural renewal," the more cynicism you'll generate.

Before you embark on an all-out war, consider a quieter approach: change by attraction rather than promotion.

Your starting point—the Preamble. Let's suppose you are in charge of a group, however small or large, and you're serious about improving meetings in your group. How do you start? With the Preamble. It's a good starting point because it lets you crystallize your thinking about basic principles—and when you're finished with it, you have a practical tool you can use almost right away. Part 1 of this series on meetings discussed the Preamble in detail and offered a general model:

> *"The primary purpose of this meeting is to improve the way we do everything here so that we, and those with whom we deal, will be more satisfied. Our guiding principles for every meeting are honesty, respect, equality, and open-mindedness."*

Just remember, the Preamble must state that primary purpose of all work done by the group. That purpose must be large enough to inspire people so they'll set aside natural differences and work as a true team. The Preamble should also summarize the principles guiding members' behavior at any meeting.

Building support

Your next task is to find a few people who'll support your ideas. You might show them this series of articles on meetings and ask for their reactions. Also, you'll want to show them your draft of the Preamble so you have something practical to discuss. When you've agreed on a Preamble, you may want to call a meeting, backed by a memo, to present your plan. Two suggestions, though: keep the memo low-key, and make attendance of the meeting voluntary.

Action

You now need just one regular meeting that can be chaired by one of the people who support your effort. If there is no such meeting, try to create one. Then print up the Preamble you and your colleagues agreed on. At the next meeting, the chairperson will simply say: "We have a new Preamble for this meeting, which I'll now read . . ."—and you're on your way.

Keeping it up

The kind of culture I am talking about will have to "grow organically." The way it does that is by encroaching on more and more meetings. For you, this means you need more chairpersons who can set the right example. Your best bet may well be a lot of informal one-on-one talks over lunch. Here are some other steps that may help:

1. Look for training that reinforces the change. Obvious areas to focus on are communication, meeting skills, and problem solving.
2. Start hiring the right people. As Robert Waterman, in The Renewal Factor, says, "pick people who reflect the values of your [group's] culture." Make a point of presenting those values at the employment interview and watch for the candidate's reactions.
3. Consider team rewards to replace or supplement individual rewards.
4. Keep an eye on the Preamble. Have people stopped reading it? Then you'll soon be back where you started. Renew enthusiasm for the "new way of meeting."

You may say, "Our boss(es) would never go for anything like this." Don't be so sure—you probably have more power than you imagine. But more on that next month.□

Primer on meetings

Part 5: Power to the people

Suppose meetings in your group or company are bruising, demoralizing experiences. You really wish something "could be done about it" (that beloved passive phrase)—but " 'they' (the managers) would never go for it." (In fact, "they" are the main problem at the meetings, you say, because of their sardonic, dominating, manipulating, or closed-minded behavior.) What can you do?

First of all, notice an interesting coincidence. When I was writing part 4 of this series, East Germans were saying just these kinds of things about their leaders. They were so sure nothing could change that a lot of them fled, leaving behind everything. Now, a month later, it seems that things can change even in East Germany. Why? Because outside conditions have changed (perestroika has become more entrenched), and because the people summoned their courage to demonstrate.

There is perestroika in Corporateland, too. Power by authority is out; power by expertise and charisma is in. I just finished reading a *Fortune* magazine article that quotes a lot of CEOs as saying so. But I can also corroborate it with my personal experience: almost all the companies we've been involved in through our teaching were going through honest efforts to build a more participative environment. It's the only way they can attract and keep top employees. In other words, you have the tide of popular change working for you. All you need is a little courage and a refresher course in the art of gentle persuasion.

Three principles

Your success in persuading the power holders to go along with some modest, gradual changes hinges on sticking to a few basic principles:

1. Get objective data. It's hard to argue with facts. And having something objective to tackle tends to keep emotions at a tolerable level.
2. As much as possible, use regular forums. You don't need an underground movement. You want open change, but without violent clashes.
3. Stay away from complaining. It's such a comfortable activity—so comfortable, in fact, that you may never get beyond it. You want to start a new, positive chapter—so move on to positive visions and actions as quickly as possible.

What are the positive things you want to remember for yourself, and ultimately transmit to others? Here are some examples:

- People have great contributions to make, to meetings and to the company. In many cases, we don't even know all their talents, knowledge, and skills.
- When the right people meet, under the right rules, they can become a true **team.** Their ideas will reinforce each other, creating a powerful "resonance" or synergy that energizes people and produces amazing results.
- Meetings can contribute to steady, long-term improvement of all areas of work. How much of that is happening now? How can we exploit this better in the future?

. . . and three steps

As I suggested last month, your initial goal should be modest. You're aiming to "invade" just one regular meeting and turn it into a success that will basically sell itself. Here is the logical path:

1. Find some colleagues who agree that meetings are poor enough to do something about it. If possible, once you have basic agreement, move the discussion to some regular forum so it will acquire official status.
2. Do a study to evaluate current meetings. Suppose you manage to enthuse four people besides you. Then there will be five people to count a few basic items in every meeting they attend. In no time you'll have an impressive information base—and you'll learn lots of useful and interesting things as you go along. Here are some simple items you might measure: how many times somebody was cut short; percentage of people who attended but contributed nothing; how many "personal attacks" occurred; how often an idea was shot down within less than one minute.
3. Present your results and recommendations in writing and orally. Be sure to include a very brief review of up-to-date literature on what a good meeting is all about, so your measurements and recommendations are grounded in authoritative theory.

The main point of the presentation is to ask for the action. And that action, at this point, is simply the blessing to proceed with an "experiment"—one regular meeting that is run on different principles, with a Preamble and the simple rules we discussed in parts 1-4. You may even want to present this experiment as just a part of your study—something to investigate and report on later.

Good luck—and I'd love to hear about your experiences.□

The secrets of public speaking

Part 1: Stamping out the no. 1 fear

Citing a national survey on human fears, *The Book of Lists* states that people's chief fear in life—outranking fear of death, disease, and bankruptcy—is the fear of speaking before a group. No doubt you know the feeling. Your name is called . . . and, suddenly, your body flies out of control. The knees refuse to straighten, the stomach rises miraculously into the throat, the heart bangs out a shattering drum solo, and laryngitis sets in with the speed—and lack—of sound. And, in this state of almost total paralysis, you have to get up in front of a sea of critical, upturned faces, smile confidently, and proceed to hold all spellbound with your suspenseful account of the history of chlorine bleaching. No wonder fear of torture isn't even a close second.

But—you don't have to be afraid. The fear that grips the potential speaker is really the terror of appearing foolish, boring, or incompetent. And the first step to vanquishing that fear is to stop thinking about how you'll look and start thinking about what you are going to say.

There are two problems to tackle here: knowledge, and expression of it. First, you must be sure you know your subject well. It may help to remind yourself that those who asked you to speak clearly believe you to be an authority; people rarely submit themselves to frozen captivity before a speaker who they feel knows no more than they do. When you think about it calmly, you will probably realize that, in fact, you know a great deal more than you had imagined in that first moment of blind panic.

Now, take out a clean sheet of paper. Write your topic across the top. Then divide the paper into four columns. Above the first, write: "What I know about this." List, underneath, the areas you feel competent to discuss. Above Column 2, write: "What I need to find out." Here, list all the important subjects within your topic about which you feel less secure. Be sure to research these as soon as possible. Even if you don't choose to talk about them, you will be ready to handle the possible question-and-answer period following your talk.

Once you have a clear view of your topic—the points you can explain and those you will soon know more about—you will be surprised how much calmer and more in control you feel! Now you are ready to turn to the expression of this knowledge.

First, consider for a moment the good speakers you have heard. One characteristic common to all is sincerity. You can tell when a speaker is trying to be funny, charming, sophisticated, or clever—and, usually, you react with a combination of embarassment and slight contempt. These are the last emotions you want to evoke! Decide, here and now, that you will not try to adopt *any* personality but the best of your own. You will add nothing to it but the crucial ingredient of all good speeches: real involvement in the subject at hand.

With this attitude in mind, let's look at the speech itself. You have three main goals:

1. to interest the audience
2. to convey a specific message about the topic at hand
3. to be sure they have understood the essence of that message.

Notice that these goals do *not* include (a) impressing your audience with your expertise or (b) amusing them. In fact, unless your speech is to be part of a cabaret performance, your audience will probably not want to be amused. Rather, they will wish to learn something and, if they're lucky, be inspired to think along new lines.

How to interest an audience? Think again of your own reactions to speakers. To be interested, you had to learn something and feel that this knowledge in some way affected you, was relevant to your own concerns. A research scientist who has been asked to talk to mill managers about his lab's latest efforts in R & D will almost certainly have something new to tell them. But, unless he or she relates this information to their interests—such as potential profitability, new markets, a technology that may help beat comptetition—they will probably come away uninterested and, therefore, uninformed.

So, we come to Column 3 on your sheet. Head it: "What will interest this audience?" Write down the possibilities, and plan to start your talk with one of them. Even if it is not part of your message, start with it. Relate your talk from the beginning to something that you know they care about. Then, when you've got their attention, they will hear what you really want to tell them.

And that brings us to Column 4. Head it: "What do I want to get across?" Under this, write one message, such as "that we need more money for research," or "that top management should pay more attention to this process developed by our competitor," or "that this project should not be scrapped." And then, list no more than three major points (drawn from Columns 1 and 2) that will convey this message.

You now have the building blocks before you, and you have only to organize them into a good speech. Fortunately, a single structure works for any type of speech, on any subject.

Step 1: Capture the audience's interest (check Column 3), by either arousing their curiosity or relating your topic to their work or lives.

Step 2: Give a *brief* overview of your topic (drawn from Columns 1 and 2) to place your listeners in the picture.

Step 3: Go straight to your message—what *you* want to say about the subject. Try to express it from the audience's point of view (check Column 3), for that is the only approach that interests them. And try to restrict your supporting points to the three listed in Column 4. With all the necessary definitions, amplifications, and examples you will have to bring to each, that is really the maximum any audience can absorb with attention.

Step 4: Restate your main points, highlights, or conclusions. Point out their urgency or significance. And stop.

By now, you are ready to "rough out" your speech. Your fears are receding, making way for new confidence and anticipation! Next time, we'll look at some specific hints for effective use of language within this spirit and structure. Remember, though, that no elegant expressions can substitute for genuine concern or enthusiasm. Or:

> Seek not for words, seek only fact and thought,
> And crowding in will come the words unsought.
>
> —Horace

The secrets of public speaking

Part 2: 'Speech finely framed delighteth the ears'

In *The Prime Ministers: From Robert Walpole to Margaret Thatcher,* George M. Thomson points out that Britain's Number One public servants were no less shattered by the terrors of public speaking than the rest of us. Before each major speech, Pitt the Younger was violently ill behind the Speaker's Chair; Lord Liverpool steadied his nerves with ether; George Canning did it with opium; Asquith, with port and/or brandy. Although these coping techniques hardly seem advisable, occasional speakers may take solace in the discovery that those who held forth every day still had a good deal of trouble doing so.

Rather than resorting to whiffs, puffs, or tipples, you can calm your nerves with knowledge. If you know what you want to say and how you plan to say it, what is there to fear? (A superior's steely glare? Look the other way!) We discussed speech preparation in the last column; now, let us consider the components of the talk itself.

A speaker's trinity

Certainly, the magic attributed to the number '3' works for speakers. Every talk built on these three principles has a solid foundation of success.

1. **Do not make more than three points in a speech.** The chairman–chief executive officer of a large corporation, who is well known for his effective speeches, has made this his cardinal rule—and it works. Whether you have been allotted 20 min to "say a few words" or an hour and a half to expatiate at length, try never to exceed the stipulated three. No one can, or wants to, take in more than three major discoveries at one sitting. Furthermore, if you tell your audience you have three points to make, you will find them waiting to catch each one. You have invested your speech with suspense—the successful storyteller's most powerful tool!

2. **Have only one thought in a sentence.** This is a good rule for all verbal communication. It is essential for effective oral delivery. Imagine your reaction if I had written:

> "This is a good rule for all verbal communication but, while it is absolutely essential for effective oral delivery, it is really not so important in letter-writing, although, surprisingly, it is crucial to the production of a good technical report."

You would probably have had to read this monster twice—if you felt like bothering. Clumsy, overstacked sentences rarely maintain a reader's interest. Now, imagine *hearing* that verbal catastrophe. You couldn't even check back to figure it out. Remember, your listeners do not have the luxury of rereading. They may not be fully attentive throughout your speech. Acoustic problems, outside distractions, and even your own lack of projection may deflect them from what you are saying. Your best tactic for making yourself understood is to make sure each sentence is clear, uncomplicated, and limited to a single thought.

3. **Follow the three-part formula,** summed up in the old adage: "First I tell 'em what I'm going to tell 'em. Then I tell 'em. Then I tell 'em what I told 'em." It's hard to beat for sheer clarity and logic. You can adhere to this "introduction—main body—summary" plan, however many variations and embellishments you add. It will keep you—and your listeners—on track.

Language of speaking

First, double your surveillance of convoluted, ungrammatical, or long-winded expressions. If you start a sentence:

"Despite the possible occasional benefits to be derived, the informal unrecognized work additions have the effect of causing. . . ."

your audience will tune out long before discovering what work additions cause—or even what they are! If you say: "In a liquid state, we have found this substance to be quite viscous," you may even elicit unwanted giggles. Be careful not to deflect an audience through confusing, pompous language or gross grammar gaffes. You may lose them forever.

Spoken language should be personal. People like a one-to-one human relationship with their speaker. They don't want to feel an encyclopedic article is being dictated to them. So, use expressions like, "I would like to share this experience with you," rather than "Experience shows that. . ." And talk about *people*—"One out of five Americans . . ."—not percentages—"20% of the nation."

Throughout, remember that everything you say will be heard for the first and last time when you say it. The audience can't go back—and neither can you. Now, ask yourself if a non-expert would follow it all with no trouble. Have you been *specific*? Could you be *even more specific*? Have you *defined specialized or technical terms* that may elude some members of your audience? Have you, in some way, restated important points? Judicious repetition can serve in speeches, precisely because audiences can't go back.

Always rehearse your speech, aloud. Where you stumble over a phrase or a speech rhythm feels awkward, change the words. A thesaurus can work wonders. Vary your sentences, alternating simple with complex and juggling the order of the parts of speech. You can be clear—without sounding boring.

One final note. Jokes. Are you a master of timing, inflection, and projection? Do your jokes relate closely to your subject and audience? Are they neither self-conscious nor insulting? Are they original? Are they really funny? If you can answer "yes" to all these questions, go ahead and use them!

Remember, unlike the terrified Prime Ministers, you are not facing a contingent that is out to tar and feather you. Your audience is interested in what you have to tell them. Say it clearly, logically, and with no pretensions—and they'll love you.

And, if the tremors still lurk in your bones, take comfort from Emerson, who declared (perhaps before giving a speech):

"The only way to conquer fear is to *do* the thing you fear."

Enemies of communication

Part 1: Fear

This enemy tends to single-out speakers, rather than writers, for its victims. Most of us would much prefer to write a memo to 100 people than to speak on the same subject to 10 of them. Why?

The "why" is important, because we have to see the enemy clearly before we can take it on. What is this universal fear of speaking before people? How does it work so powerfully on so many? How does it affect your efforts at oral communication?

A theory now in vogue says that the fear of public speaking stems from our herd instinct. We're all right so long as we're part of the pack, with people all around us, literally to back us up. But take one of us out of the herd and isolate him in front of the others — and he's terrified.

The "herd" theory may have some basis, but if so, it seems to be relevant *only* to public speaking. Most of us are happy to do our work alone, with no one standing nearby to support us. Indeed, we crave space and even solitude in which to work. Why, then, should we be afraid to leave the herd only to speak?

Fear and illusion

I believe the fear of public speaking is more specific — and hence more manageable. It's manageable because it can be clearly defined and then unmasked, for it's a fear based on an illusion.

The fear of public speaking is a composite fear of:

- looking gauche, unsophisticated, clumsy, or anything else that counteracts the image you want to project
- sounding shrill, gravelly, squeaky, tongue-tied, or anything else that jars with the voice you want to be yours
- giving an overall impression of being boring, ridiculuous, incompetent, unprepared, or unintelligent
- if speaking from notes — losing your place or dropping pages, papers, notes, pointers, chalk, or other props
- if speaking impromptu — having nothing to say, having nothing of importance to say, or forgetting what you did have to say.

Now, this looks like a pretty big, multidimensional fear to crack. But it isn't really, you know. Because all these five parts are symptoms of a single fundamental error — an illusion. Once the error is unmasked, the fears dissolve.

The error unmasked

The error made by the fearful speaker is this: *he thinks he has to make a particular impression on his listeners.*

He is wrong. The audience isn't looking to him to make an impression. People are there to hear what he has to say. He is there to *tell somebody something*. Period.

Think back to the speakers who have made their mark on your life. I am reflecting on one famous public speaker I had the privilege to hear a few days ago. He spoke in a gravelly voice which sometimes interfered with the clarity of his diction. He was short, balding, and not particularly impressive to look at. Once, moving across the podium, he tripped. When he had spoken for about two-thirds the allotted time, he apologized for having devoted too much time to his first point, leaving less for his second. *He did just about everything we're all afraid of doing!* And he was wonderful. The audience walked out spellbound. Why?

Because the man wasn't concerned with the impression he was making on us. He was there, vibrantly there, to tell us something that he thought was very important. He was, if you like, the servant of an idea which he thought worthy of all his energy and urgency. And so, he was wonderful.

What you say must matter

Now look back at that analysis of the five-fold fear of public speaking. Does one omission strike you? The omission is: the subject of the talk. If you're concentrating on your appearance, your voice, your image, the impression you make or the reactions you produce — you really don't have a moment or a thought left for the subject of your talk! And do you know what happens, when you're tied up in all these fears? As Job lamented, "The thing which I greatly feared is come upon me." You leave exactly the impression you were so afraid of creating.

Why? Because the only way to give a talk is to concentrate on getting your whole message across to your listeners, in terms they can understand. We'll talk about ways to do this in another column. Now, just hold onto it. By "whole message," I mean both the facts and their meaning and importance to you and your audience. Focus on communicating this whole message — and your fears about your appearance or your "image" will disappear.

By definition, we can concentrate on only one thing at a time. Once you start concentrating on the substance of your talk, you can't worry about the impression you'll make. And the magic is: as your fears about your image evaporate, *you will look good.* You may drop the chalk. You may even trip trying to pick it up. You may speak in less than mellifluous tones or stand in a less than glorious pose. It doesn't matter. Your listeners will see and hear you as terrific. Because you'll be giving them what they want: a message to think about. That is all they will focus on, too.

You can do deep breathing and relaxation exercises before you speak, if it makes you feel better. My feeling is that such pyrotechnics make sense only if you always do them before talking to some friends. That's what you'll be doing, when you get up to speak next time. You'll be talking to a group of decent human beings, people like you, about something that interests all of you.

Nothing frightening in that, is there? □

Enemies of Communication

Part 2: Facts, facts, facts

A departmental task force had been created to study the advisability of using a large computer to store certain research-related records. Although only barely informed about computers, the members of the task force studied their possible purchase most diligently. Their investigations led to a unanimous conclusion: the computer was essential. It could save a great deal of time and cut down on losses and errors. But they saved that conclusion for the end. Their report to top management began with an exhaustive description of the special features of this new machine:

> The data retrieval system of the XYZ machine allows for reel re-use within a single shift. This obviates the necessity of creating storage space for temporary placement of used reels or of logging in the used reels prior to changing their destination. With this unique feature, there is also no need for a library of volumes, backup or scratch tapes, or an elaborate filing system for the reels. Furthermore . . . .

Few people finished reading the report.

An assistant research director of a paper company had been assigned to take some new members of the sales team on a tour of his department's lab. One young rep asked him how the brightness of a particular sheet compared with others on the market. The assistant director extolled his company's product, declared it superior to all other brands but one, and then treated his captive audience to a 25-minute discursion on the different processes of brightening mechanical, semichemical, and chemical pulps. No further questions were asked.

Both the writer of the report and the speaker to the group were eager to communicate, well-informed, and enthusiastic about the subject. But both would-be communicators lost their audiences. Why?

The answer is simply: too many facts. The writer and the speaker believed, mistakenly, that an outpouring of facts equals communication. On the contrary — it usually destroys it.

The "whole message"

In the last column, on public speaking, I talked about the whole message. I promised that you would succeed as a speaker if you concentrated purely on getting your whole message across to your listeners. The whole message consists of:

- facts that are important to your audience,
- explanations that your audience needs in order to understand those facts,
- your conclusions or interpretations of those facts.

The facts are only part of the message; and the message contains only certain facts.

Here is a vital law of communication: *The facts are never the whole message.* Paste it up somewhere, either in your office or, figuratively, in your mind.

Finding the whole message

The eager task force and the assistant research director were ignorant of that law. Having learned all about the wonderful world of data processing, the task force felt compelled to share this new-found knowledge with its readers. But most of the readers neither understood the terms used nor felt a need to know about all the workings of the XYZ machine. They wanted to know, in plain English, whether the machine would help them store and retrieve particular information. The members of the task force should have asked themselves exactly what benefits the company might hope to derive from the computer and what hazards the machine might introduce. For instance — could just any user tap the records? How could restricted records be kept restricted? Such facts as these are the ones the readers would seek.

Similarly, the assistant research director thought he ought to tell his inquirer everything he knew about brightness. Instead, he should have asked himself what the new sales representatives would find useful for their jobs. What facts might be important to them? And how could he make those facts understandable, in non-technical terms? Keeping these questions in mind, he would have answered the question on brightness differently. He would have realized that the person asking the question wanted to be able to use the answer to bolster his sales pitch to potential customers. So, he would have told him that the company's product had a brightness of X%, compared with certain other companies' level of Y%. He might have added that the competitor with brighter sheets used a more costly method of bleaching — costs that were passed on to the customer — or that the competitor's bright sheets were actually much weaker than the ones the reps were now viewing. He would have communicated successfully with his listener by telling him the facts *he* needed, the facts that were important to *him*.

Next time you speak or write about a subject that you have studied in some depth, ask yourself three questions:

1. Who are my readers or listeners?
2. What facts do they want to know about this?
3. In what order?

Give them only those facts, in that order. If they need some explanation in order to understand the facts, supply it. *But don't go beyond what they need.*

Remember, the whole message consists of the *facts* your audience can and wants to understand . . . *explanation* of any of these facts that may be unclear . . . and your *interpretations* or *conclusions*. Then your facts will serve you and your readers.□

How to give technical presentations without hating it

One big reason most people hate giving presentations is fear of boredom. As one engineer put it: "I'm always afraid they'll fall asleep—and they usually DO."

Why are most technical presentations so unbearably dull? Because they hold no surprises. The audience gets exactly what it expected—usually, an endless series of "word chart" transparencies, accompanied by feeble poking at the screen, and almost verbatim repetition of what everybody can read there. You need two spices for this dish: SURPRISE and VARIETY. These are the elements that wake up the audience—and they're also the ingredients that lead you very naturally to a very lively, powerful delivery.

How to put surprise into your talks

The biggest surprise to your audience will be a clear, exciting main message, stated up front.

Most speakers start with topics rather than messages—say: "Our objective today is to review the progress of Project X" or "I'd like to explain our company's purchasing policies and the reasons behind them." But if you're just announcing a topic, you're not saying anything about it—and if you're not saying anything, how exciting can you be?

Contrast this with true messages like "Project X has been a stunning success so far, but it will fall behind schedule unless we commit more people and money to it right now." Or: "By cooperating with the purchasing department, you can get better deals and avoid all kinds of contract-related-headaches—and if you follow a few simple points, you won't have delays or a lot of red tape, either." Here, the speaker is shocking the audience with a pithy statement that can stir hopes and fears.

Note the phrase "that can stir hopes and fears" in that last sentence. This is what separates exciting messages from the "ho-hum" variety. You're one step ahead if you find a clear main message at all. But to make it exciting, you need to shape it so it touches your listeners' basic needs. So think about your listeners, and test your main message against their needs and interests. What do they want when it comes to purchasing big or small items for their work, for example? They might want speed, choice, low cost, no red tape, and favorable contracts. Address those points, and you're giving them exciting news; talk about the history of purchasing procedures and how they make YOUR work easier, and you'll find them snoring in no time.

The professional touch: variety

What keeps us awake is change of any kind. This is why polished presenters will constantly change tactics. You can do the same—but you have to PLAN for it. Here's how:

Choose about three (at most five) key points that expand your main message. Then illustrate or back up each point, consciously switching method. There are more tools available to you than you might imagine:

- Specific examples of a process, project, or idea
- Anecdote or story
- Analogy to explain a concept or give life to numbers
- Quantitative trends, expressed in a bar chart or trend curve
- Quote (especially from an authoritative source)
- Demonstration (holding up or showing something that illustrates your point)
- Color slide showing people, equipment, buildings, etc.
- Short videotape showing an operation, interaction, response, or dramatized "incident"
- Flip chart (e.g., for recording audience response)
- Audience involvement through simple questions
- Simple graphics or cartoons, using overhead transparencies or slides
- Drawings of equipment or processes
- Short exercises, puzzle, or a group game that helps the audience appreciate a point.

Consciously planning for a variety means you will never put on one transparency after another, without even switching off the projector. Instead, you'll show one graphically interesting transparency, turn off the equipment, and then switch to a personal example, quote, or other tool. You'll have achieved variety—and you'll convey power and control, because YOU run the equipment, not the other way around.

Freeing yourself from the slavery of the overhead projector has another big benefit: natural variety of MOVEMENT. While you're working with a screen image, you'll naturally move to the screen, touch parts of the image, and bring the audience "into the picture" with the other hand. Once you switch off the projector, you can move to the other side of the room without creating annoying shadows. And when you're ready for another visual, you can calmly place the transparency on the projector while finishing your point. Then switch it on to get an instant, perfect image without any fumbling.

A presentation gives you a unique chance to sell your ideas and yourself. By systematically planning for surprise and variety, you'll make the sale and have fun along the way! □

A universal speech plan

I hope you've had the opportunity to try the Universal Outline I gave in my last column. It will help you write more quickly and effectively.

Now I have a similar tool for you for public speaking: A Universal Speech Plan.

Whenever you give a talk, you have just one goal: To give people something of value and show them how to use it.

This is not the goal most speakers set for themselves. Rather, most seem to have one of three different aims:

1. Show them how smart and in-the-know I am
2. Get the audience to do what I want
3. Get through this terrifying ordeal somehow.

No wonder so many presentations are boring and so many speakers scared of giving them. These goals don't work.

Speeches that work

Think about yourself as a member of the audience. Are you there to watch someone prove how clever he is? Or to hear what he wants you to do for him? Or to suffer for him as he endures chronic paroxysms of stage fright? Of course not!

What you want as a member of the audience is a speech that gives you something of value and shows you what to do with it. A speaker who does that for his audience will always succeed—no matter what technical or vocal glitches get in his way.

You'll read everywhere that a speech consists of three parts: Introduction, Body, and Conclusion. This is indeed true—but to leave it at that is to miss the point. These three parts correspond to something more vital: they are the three phases of giving the audience something valuable.

That's how you should always think about your speech or presentation: three stages of giving your audience something valuable. First, you greet them. Then, you give them something worthwhile, with an explanation of its significance to them. Finally, you provide directions for its use.

The universal speech plan

Here is my Universal Plan for constructing an effective presentation around the concept of giving something of value.

Introduction: Greet your audience

Establish rapport through a statement that connects you to the people you're addressing. Tell them briefly what your main message is and why it's valuable to them. If your presentation is fairly long, give them a short outline of what you'll cover.

Body: Give them something of value

Substantiate your main message with three key points. Mention all three. Then take each one, stating the principle or fact, explaining briefly, and giving an illustration or example.

Conclusion: Show them how to use what you've given them

End with a strong example or illustration of your message and a statement of what the audience should do to benefit from it.

Here's an example

The press section of one of your company's paper machines is causing problems. Your are to give a presentation to the firm's Capital Committee to recommend buying an extended nip press to take the place of the plain press now in use.

Introduction

Thank the committee for their time and tell them briefly how your problems relate to company welfare. Explain that the antiquated press equipment is costing the firm time, money, and quality. Say you recommend replacing the plain press with the extended nip press. State that you will explain how the new press will solve the problems, improve quality, and save the firm time and money.

Body

Summarize the three problems—e.g., irregular caliper, crushing of web, paper breaks. Then take up each, showing how the present press configuration contributes to the problem, what is needed to solve the problem, and how the extended nip press would do it.

Conclusion

Show a chart illustrating the cost-effectiveness of improving efficiency in the press section, pointing out that water removal is pressing costs only one-tenth of what is would cost in the drying operation. Point out once again that the extended nip press would give the company a better product for less money. Thank them and sit down.

You can use this Universal Plan for any talk or presentation you give. Remember, a presentation is just that: something you give. This plan merely follows the natural pattern of giving someone something of value and showing him how to use it. Good luck!

Chapter 8
The Key to Quality

How to Make Communication a Quality-Improvement Tool

This final chapter presents some of the extraordinary benefits of good communication at work. Improved communication can be your number one quality improvement tool. It can yield immediate results. Conversely, without good communication, a whole barrel of other expensive "quality" tools can be simply an extravagant waste of time.

Why bother with communication? lays out the great and very specific benefits you can derive from improving communication in your group. **Communication—a key to quality control** is a five-part series that provides you with five solid techniques for improving quality through skilled communication.

To think clearly—write clearly demonstrates how sharpening your writing skills can actually sharpen your *thinking* skills. **Writing to solve problems,** a three-part series, gives you three ways to use writing as a problem-solving tool.

Managing writing, a four-part series, shows managers how to improve both the writing and the morale of their groups.

And the last series of columns, **Principles from poetry,** demonstrates how reading poetry can help you write more powerfully and succinctly, exploit the power of the unexpected, and think in new and wholly creative ways.

Why bother with communication?

You're a technical manager—a research director—a VP Operations—an engineer. You're not at work to talk to people. Except for the occasional brief memo or technical report, you don't write that much. Why should you spend a moment of your busy life thinking about communication, of all things? You're there to get the job done well. That's enough of an effort.

Why bother with communication? Very simply, because you can't get the job done well without it.

Here, I will show you a few reasons to take communication seriously–seriously enough to make the terrific effort required to change behaviors of a lifetime. In following columns, I'll show you how to begin to make those changes.

To be an effective professional or executive, you must be able to communicate

In *The Effective Executive*, Peter Drucker calls executives and other professionals "knowledge workers"—people who produce not physical objects but rather knowledge, ideas, information. Then he make the crucial point:

By themselves, these "products" are useless. Somebody else. . . has to take them as his input and convert them into his output before they have any reality. . . knowledge work is defined by its results.

Knowledge, imagination, and intelligence are not enough. To be effective–that is, to get the job done well—you must be able to communicate your knowledge and ideas to the people who can convert them into the output or product that the department or company needs.

To create an effective product, you must be able to communicate

How do you know what your company or department needs to produce? If you're thinking, "Give me a break, that's obvious" —think again. This question touches on one of the most widespread problems I've encountered in my communications workshops in companies of different sizes and from different industries. Mill people complain that those in the technical center don't produce the findings they need. Conversely, researchers say the people at the mill don't give them the information they need to do the necessary research. Sales representatives complain that they're not getting products that meet the customers' requests. And customers complain that the companies' product does not live up to its promises.

An effective product is one that works and fills needs. To create an effective product, you must be able to communicate, across boundaries. Find out what your staff wants and needs from you—that's your effective product. Find out what the other departments want and need from yours—that's your department's effective product. Find out what customers want and need—that's your company's effective product.

To run an effective group, you must be able to communicate

You must be able to give your staff clear and understandable instructions. If they don't know exactly what you want them to do, you'll have chaos on your hands. Clarity is essential. But it's not enough.

Do you want your people to work as hard as they can, with all the enthusiasm and creativity they possess? Giving them clear, complete information won't do it. You must be able to inspire them to put out their maximum effort—with pleasure.

In a recent interview with *FORTUNE* (Aug. 9, 1988), Lee Iacocca talked about one of Chrysler's most successful plants:

. . .it appears that our plant with the least automation and least investment is turning out the best quality. You ask, "What's going on?" Well, it has to be the people in the plant, management and labor. It is how the people approach their jobs that does it. . . .The answer is to make a guy—any guy—feel that when he comes to work he does something, and contributes something, so that he can't wait to come back tomorrow. . . .

This, to me, is the essence of effective communication at work. "To make a guy—any guy—feel that when he comes to work, he does something, he contributes something, so that he can't wait to come back tomorrow."

Communication is a central, essential skill for you in your work. Once you realize that you cannot perform effectively without good communication skills, you'll start to pay attention to what you're doing now and what you could do better. That's the first step.

Communication: A key to quality control

Part 1: How to ask questions

At a seminar on communication that I gave to TAPPI's Engineering Management Committee, I asked participants what they thought was the biggest communication problem on the job. One manager immediately responded: "People don't ask enough questions."

According to an engineering manager, "People don't ask questions about a job they are working on for fear that the questions would reveal their ignorance of at least some part of the work. Instead, they follow the simple principle of shooting in the dark in the hope that, eventually, they'll hit the target." Predictably, the results verge on disaster.

"If they would only ask a few questions at the beginning, they could save a whole lot of botched jobs," the engineering manager lamented.

Indeed, asking questions is a *critical* first step to quality control. If you know precisely what you are supposed to be doing and what you need to do it, chances are much greater that you'll do it the right way.

Why don't we ask questions?

Obviously, says the cool voice of reason, it makes great sense to ask questions when you're confused about something important for which you are responsible. After all, you're better off admitting your ignorance in the beginning than having it blow up in your face (sometimes quite literally) at the end. Who ever said human beings listen to the cool voice of reason?

No—our self-protective fear chimes in—don't show yourself up. You'll figure it out along the way. Just use your common sense. Finesse it for now; later, you can look it up. Or talk to Joe, maybe get him to drop a few hints. Or—something. *Just don't admit you don't know.*

Terror drowns out reason, and we go on smilingly and silently to botch the job.

The key to asking questions: humility before the job

One quality of greatness appears to be great humility before the task at hand. The best among us don't stop to say, "How will I look if I do or say this?" Rather, they are totally focused on doing the job superbly, to the very best of their ability and to do so, they know they have to ask questions. Lots of them.

Albert Einstein once said, "Nothing was ever obvious to me." So, he asked questions.

We can all take immediate steps to improving the quality of our work by determining to ask effective questions.

Type 1: questions to get information to do a job well

There are many types of questions, some of which I will discuss in upcoming columns. Here, I would like you to consider the most straightforward type: questions to get information to do a job well.

Questions to get information to do a job well

Who . . .

- Is involved?
- Is doing what here?

What . . .

- Do I need to do my part?
- Exactly are we trying to accomplish?

Where . . .

- Will it be done?
- Are the people I may need to contact?

When . . .

- Must my part be completed?
- Must the whole job be completed?
- Can I reach the people I may need?

How . . .

- Do you want it done?

Before you undertake any project of significance (I suspect Einstein would say that includes every project), check that you know the answers to all those questions. If you don't, ask the appropriate person.

By going systematically through the Type 1 questions, you will also see what you need to know. Part of the problem is often that we don't realize at the beginning just what we need to know to do a job completely and efficiently.

From my own experience and that of many others, I can assure you that these few questions will guide you to the information you need.

Of course, that's not all there is to it. You have to be able to *listen effectively* in order to process and use the information you receive. We'll come to that later in the series.

Before you start your next project, take the time to check all the "questions to get information to do a job well." Let me know how it works for you!□

Communication: A key to quality control

Part 2: Managers—think before you edit

"He who writes carelessly confesses thereby at the very outset that he does not attach much importance to his own thoughts . . . A man convinced of the truth and importance of his thoughts feels the enthusiasm necessary for an untiring and assiduous effort to find the clearest, finest, strongest expression for them . . ."

—*Arthur Schopenhauer On Style*

Each person's writing is an expression of his or her thoughts. We all know that. Right? Well—if we do know it, we certainly don't behave as if we do. And the results of this neglect are deeply damaging to motivation and, by extension, high-quality work.

Let me explain.

As most of you know, I teach in-house writing courses to corporations. A recent comment by a research scientist in a paper company crystallized a concern I had been hearing over and over in the classes:

"I don't understand why the company sent me to this course. They're not interested in what I write."

When I asked him to elaborate, he went on to tell me that his manager edits everything he writes so heavily that nothing of his original work remains. He feels resentful about this—but, even worse, he feels that his thoughts are considered unimportant.

Can you imagine what this feeling does to this scientist's motivation? Yet I hear the comment over and over again:

"My boss edits me out of my reports."

Remember: writing is the expression of a person's thoughts.

I went to see the manager of the research scientist and told him that this man felt somehow undervalued because everything he wrote was so heavily edited. The manager expressed surprise at this reaction. It turned out that he considered the scientist quite a good writer; his only complaint was that the man "wrote too much."

He added that the man was "one of our better scientists."

"Then you do value his thoughts?" I pressed.

"Of course."

"You wouldn't think of reaching into his brain and twisting them around to fit another pattern?"

"Of course not. What sort of question is that?" He was starting to look at me strangely.

"A reasonable one," I said, playing my trump card. "Because that's what you're doing when you edit his expressions of his ideas out of his writing." Bringing out the heavy artillery, I quoted Schopenhauer to him. I asked him how he would feel if he had carefully written down his interpretation of an event, only to have it summarily dismissed with a slash of red pencil.

He nodded. "I understand. But—he does write too much."

I told him we would be working on tightening up the man's writing in the writing course. But I asked him to consider one question. Which was more important: to show this bright man respect for his ideas, thereby encouraging him to do his best work—or to cut his three pages down to two? The manager understood.

A company's greatest "quality asset:" its people's thoughts

When we think of "quality" in the workplace, we tend to think of objects (the products and their components) and actions (the work done on the job). There is a lot of talk about improving the quality of service and production and teamwork. But these actions, critical as they are, are not the roots of the company effort. *Thoughts*—ideas, plans, visions, solutions to problems—are the source of the actions, which in turn lead to the product itself.

To improve the quality of our products and services, we must go back to their roots: the thinking that developed them. We must foster creative, problem-solving thinking on the job. One way to do this is to respect people's expressions of their thoughts.

Managers—hold the red pencil

Of course you have to edit your people's reports. But the editing you do should be strictly limited as much as possible to correcting errors. Fix mistakes in content or grammar. Then stop. Don't slash through the piece. If it is wordy or confusing or badly organized, please *resist* the temptation to delete paragraphs or rewrite the whole thing. Instead, make suggestions to the author. Circle areas you think are unnecessary and ask if they might be dropped. Ask for clarification of a confusing passage. Underline points you think are important and suggest that the author move them up.

Remember, you're dealing with one person's most sensitive and important asset to the company: his or her thoughts. And thoughts don't take kindly to mutilating, even on paper.□

Communication: A key to quality control

Part 3: Listen for the message

A few months ago, an Avianca jet crashed not far from where I live. Among those killed were relatives of two people I know. They had been on their way to New York for a family visit.

It appears that the plane crashed because it ran out of fuel. However, inquiries into the disaster suggest that the pilot had told the people in air traffic control that his fuel was running out. He did not, however, use the precise words that indicated **"emergency."** Hence, a space was not immediately cleared for him to land.

The cause of all those deaths may have been a problem with communication.

For people—effective listening is critical

If the pilot had been addressing a computer on the ground, there would have been no question of the fate of that plane. A computer can react only to very specific commands; if it doesn't get the precise signal for "emergency—clear runway," it will not clear the runway. It's that simple.

But unlike computers, people can go beyond words. They can listen and react to the message behind the words. When they don't, the results can be tragic.

In the previous two columns on communication and quality, I discussed the importance of asking thoughtful questions (*Tappi Journal,* July 1990) and doing thoughtful editing (*Tappi Journal,* August 1990). In both cases, the first step is stopping to think.

Before you start on a project—stop to think. Have you prepared yourself completely by asking the questions necessary to get the information you need?

Before you edit someone's report—stop to think. Have you respected the writer by limiting yourself to correcting errors and making suggestions?

Now, as we move to effective listening, the first step is the same. Before you react—stop to think. Have you understood the message behind the words? Do you know what the speaker wants from you?

Thoughtful, attentive listening is hard work. It doesn't come naturally. Once we think we've understood, we tend to "tune out" and get on with something else: airing our own views, thinking about something that interests us more, taking the action we assume is required, or—not taking the action.

To undertake the regular self-discipline of good listening, you must be convinced that listening for the message is worth your trouble. I hope the example I gave will make you consider the importance of careful listening. But before you go on reading, take a moment to reflect on a company project or effort that has had problems. Can you think of a situation that might not have happened if someone had listened more completely? It shouldn't take you long; poor listening is one of the most common causes of errors and, hence, poor quality. Sometimes it leads to disaster. Many suicides have been quoted as saying, "Nobody listened."

How to listen

Try this four-step method of listening. It works.

1. Give the speaker your *full* attention. Look directly at him/her. If you're talking on the phone, try to imagine him talking to you.
2. Clear your mind of all other thoughts. Keep on the watch for them as they try to intrude, and refuse them entry.
3. As the speaker talks, keep checking your understanding. Do you understand what he is trying to get across? The words themselves rarely give the whole message. Listen for the tone. Is the speaker upset, angry, worried, excited, resentful, hesitant, scared? Stop the speaker every so often to see if you understand what he's really saying. You can check your understanding by asking simple, direct questions: "Do you mean . . .?" or "Are you concerned about . . .?"
4. Try to determine what the speaker needs from you. Ask yourself: "Why is he telling me this?" Sometimes, as in the case of the Avianca plane, a question would suffice: "What do you want me to do?" Other situations require different responses—as we'll see next time, in a column on the three main types of listening.

Listening is a vital, terribly ignored tool of communication. If you know of a case in which listening caused or solved a problem, please send it to me. I'll print it for the benefit of your colleagues.□

Communication: A key to quality control

Part 4: The three types of listening

"Can you believe it?" my friend exploded. "My boss left the entire report for me to do—and then just took off on vacation. He never even discussed it with me; all I had was a bunch of unreadable notes on top of a pile of papers. Of course, he doesn't care; he knows who'll take the rap if it gets messed up. Well, maybe when he gets back from the Bahamas, he'll just find the same pile of papers on his desk instead of a report. That might teach him how to treat people."

Alarmed at my friend's self-destructive course, I tried to refocus him on the task of getting the report done. "Is there someone you can go to for more specifics on the report?" I asked him. "No! Why should I run around begging for information?" Perhaps I needed to be more direct. "When is the report due?" I demanded. "The first of the month. Hah! That'll show him. He'll get back with only a week to put it all together."

This went on, with only a change in intensity. The more urgently I tried to pull my friend back on track, the more vehemently he insisted that he had to "show" his insensitive boss. Frustrated, I told him that he was setting himself up for being fired. And then he said the all-revealing words: "You don't understand." And at last, I did. I had been doing the wrong kind of listening.

Learn to listen—listen to learn

My friend didn't want advice or guidance. He wanted **emotional support**—and the more I withheld it, the louder his grievances became. I was contributing to a real quality breakdown, as I unwittingly entrenched my friend in his determination not to write the report.

Phrase Three of the **Four Phrases of Listening** [Tappi J. 73(9): 332 (1990)] tells you to ask yourself: "Why is he telling me this?"

That's where I fell down. Worried about my friend, I just reacted. Had I asked myself that important question, I would have realized that he wanted my emotional support. When I finally gave him that support instead of the unwanted suggestions, his rage began to diffuse. Then he was ready to start thinking about ways to write the report.

There are three main types of listening. Your answer to the question "Why is he telling me this?" will show you which type to use.

The three types of listening: active, supportive, and creative

Active listening: showing you have thoroughly understood the information imparted

When to use:

1. The speaker talks calmly to give **information**
2. He **doesn't ask** for advice or involvement
3. He uses **descriptive** words, such as: "It works like this . . . ," "The next step is . . . ," etc.

What to do:

1. Listen closely to get the **essence** of what he says at each stage.
2. Periodically **restate** what you think he said to make sure you've understood. ("Let me see if I've understood up to here." "Do you mean . . .?")

Supportive listening: showing you empathize with the speaker's feelings

When to use:

1. The speaker is **agitated.**
2. He uses **emotional** expressions, such as: "Can you believe it?" "I told him that was it."

What to do:

1. Give complete attention to the **person**, not the subject.
2. Think throughout: "I'm just going to listen and show I understand and **support** him."
3. Periodically **reflect the speaker's emotions**, with support: "You must feel really hard pressed . . ."

Creative listening: building on the speaker's ideas to form something new: a new idea, a new application, a new solution to a problem

When to use:

1. The speaker says calmly that he wants to **discuss** a problem or situation.
2. He **invites a response**: "What's your opinion on . . . I'd like your thoughts on . . ."

What to do:

1. Listen for the **main points.**
2. Periodically test your understanding (see Active Listening) and ask **Who? What? Where? When? Why? How?** questions to help the speaker clarify thoughts.
3. Pick a point and try to **add** something to it: "Let's think about other ways to get management to support buying the computers. Could people use the portables on field trips? That would add a new use and improve efficiency."

Effective listening drives communication on. If a speaker stalls on a single topic or emotion, he probably isn't getting the response he needs. To move a conversation on to useful results—try listening differently.☐

Communication: A key to quality control

Part 5: How to make your writing a quality improvement tool

How can you use your writing to improve quality on the job? The answer is simple: tell your readers clearly and simply what they want to know.

The answer may be simple, but as most letters, memos, and reports prove, the execution of it is extremely difficult. Telling your readers what they want to know is wrenching work, when you're burning to write about what interests you.

But if you do manage to follow this single guideline—every bit of it—you will improve the quality of work, as you contribute important information effectively to those who need it to do the job well.

To see how to put it into practice, let's consider each part of this guideline.

Tell . . .

Before you put pen to paper or finger to keyboard, think: "I want to *tell* X that . . ." Do **not** think: "I want to *write* about . . ." The word *tell* is critical. It guides you to the straightforward, honest language of everyday speech.

If you start thinking *write* instead of *tell*, you'll be surprised how your words suddenly mutate and multiply. *Here is* . . . explodes into *As per your request, enclosed please find* . . . *When* expands into *at such time as*, *if* into *should the situation arise that*, and so on.

Even if you avoid the puffy polysyllabic, your attitude changes when you slip into the "write" mode. You begin to think: "How shall I write this?" instead of "What do I want to tell this person (or people)?" And before you know it, you're wondering about words and phrases instead of thinking about the message that you have to transmit.

. . . your readers . . .

Tell . . . led you to the right *words* and *attitude*. The next two words point you to the right *focus*: your readers. Before you start writing, imagine who your readers are and what their interest in your topic might be. Then you'll have a good beginning idea of what to tell them.

. . . clearly and simply . . .

This is your guideline for *style*. Good style is not one that sounds "businesslike" or "formal" or "technical" or even "professional." Good style in business writing actually doesn't sound at all. You don't notice it, because it is totally in service of the message. It is clear and simple.

If you're focused on telling your reader what he wants to know, you will naturally search for the words that tell him as **clearly** and **simply** as possible.

. . . what they want to know . . .

This final part of the guideline gives you the **content** of an effective piece of writing. To be sure you write something that people will read, answer three questions:

1. What does the reader . . . or, if several, the primary reader . . . **already** know about the subject?
2. What else would **he want to know** about it?
3. What do I want to tell him?

Your answer to Question 1 will show you what you do and don't need to cover. Why waste your reader's time telling him what he already knows? Just give him enough background to explain and clarify your message.

Then answer Question 2 right away, as close to the beginning of your piece as possible. Tell your reader right up front what he wants to know.

If you're in luck, the answers to Questions 2 and 3 will be the same. Sometimes fortune smiles on us, and to our delight, we find that what the reader wants to know is actually what we are burning to tell him.

But frequently, this is not the case. Hence the writer's dilemma: Which do I tell him first? Do I start with the vitally important point that I absolutely *have* to tell him? Or do I begin with the somewhat ancillary and frankly trivial thing that interests him? What if he reads that and stops there?

The answer is that busy people read only those things that they consider important. If you start with the point that interests your reader, you will get his attention. He will also feel positive about you and be willing to go on reading.

Start with his interest. Then go on to your interest. *But be sure to show him how that second point—your interest—* **affects him**. If you don't—he won't read it.

If you're not convinced, try stepping into the reader's place. Which would you like to find up front: the point that interests you or the one that interests the writer?

Tell your readers clearly and simply what they want to know. That's all. Simple, isn't it?□

To think clearly—write clearly!

This letter came from the customer service department of a major credit card company:

"The following is in explanation of the temporary credit of $633.11 which appeared on your November statement. This credit was awarded erroneously to your account referring back to the equivalent debit which appeared on your June statement. Since your account was originally credited in September for the error which appeared in June, our cashier's department has now resolved the outstanding discrepancy by reversing the temporary credit awarded you in error and we have closed our file.

You will be glad to know that your account now shows a balance of zero for this transaction."

Who in his right mind would be glad to know that his account showed a balance of zero? But besides that glaring error in human psychology, the writer made a few others.

If his first few words are to be believed, he did *not* intend to baffle his reader or reduce him to a heap of helpless giggles. He wanted to *explain* something to him.

What happened?

I believe the letter-writer simply recorded his thoughts, as they occurred to him. This is how I would reconstruct his thought process:

"I have to tell X why a credit of $633.11 showed up on his November statement. Let's see—it was a mistake. We gave him credit for something we'd charged to his account in June. Why did we do that? Oh, I see. We billed him for a purchase he never made. But we'd already given him credit for it in September. Then some idiot gave him credit *again* in November! OK. So now we took away the second credit, and everything's hunky dory."

The order of thoughts is *not* chaotic, but, as you can see from the letter, its written translation is practically incomprehensible to all but the writer himself. That's one of the biggest problems with confusing explanations. The writer doesn't see them as confusing. To him, they look crystal clear. He has put his thoughts on paper, as they came to him. As he reads it, the written translation of those thoughts makes perfect sense.

Seeing through the mud

How can you avoid writing something that may be lustrous crystal to you but mud to your readers?

First, you follow a simple system of logic. Whenever you want to explain something, proceed by answering these three questions, in this order:

A. What *fact* or *event* am I trying to explain?

B. What *led* to A?

C. What are we *doing* about A?

Then, follow these rules of good writing as you answer the questions:

1. Imagine that you are *talking* to your reader.

2. Set down events in the order in which they occurred.

3. Write only one event or fact per sentence.

Now let's see what happens to that mind-bending letter when we subject it to the three questions and rules of good writing.

A—What fact or event am I trying to explain? I am writing to explain the credit of $633.11 that appeared on your November statement. The credit was our mistake. Please let me explain what happened.

B—What led to A? As you know, a charge of $633.11 appeared on your June statement. This charge was a billing error. We corrected it on your September statement by giving you credit for the same amount. That closed the transaction. Nevertheless, our cashier's department mistakenly repeated the credit on your November statement.

C—What are we doing about A? We have now removed the extra credit from your account. Please forgive us for this confusion. Thank you.

George Orwell said, "Slovenliness of language makes for foolish thoughts" (*Politics & the English Language*). The reverse is also true. If you demand clarity and logic of your language, you will *impose* clarity and logic on your thoughts. You'll be repatterning your own mental processes, from the outside in!

Writing to solve problems

Part 1: Writing to discover

In our writing workshops, we sometimes ask participants to write and hand in two short descriptions—one of a person, the other of a project. Here is a typical response:

1. My mother is small, with silver-gray hair and blue eyes. Although her hair is basically white, she doesn't really look old. She has a soft voice and gracious manner.

2. A project I'm working on right now is a task force designed to establish accurate job descriptions for the people working in our department. First, we designed a questionnaire to send to people. Now, we are interviewing them one by one.

Then, we ask the class to write about the same subjects—this time not for us but for themselves. Their goal is *to discover what they know or feel to be important* about the subject. They don't have to show this writing to anyone, but some students have been good enough to share their two sets with us. Here is a second set written by the author of the paragraphs quoted above:

1. My mother is strong beyond anything I could ever be. That tiny body, so frail and in such constant pain—but she hardly ever complains. She keeps her hair done and her face perfectly made up to show the world a pretty picture that challenges the pain inside.

2. This so-called task force project is destroying the department. No one trusts anyone anymore. When I go in to interview someone, I know he's lying to me about what he's doing and resents me for poking into his affairs. And I don't blame him.

Writing to communicate vs. writing to discover

When this student was writing the first descriptions, he was constantly aware that someone else—a stranger—was going to read his words. Therefore, he is writing with severe constraints. He wanted not only to describe the person and the project but also to limit what he revealed about them. Whether or not he realized it, a primary goal was to give the reader a certain impression.

In the first example, he writes about the physical appearance of his mother, to keep the description objective and on the surface. But notice how quickly he tells the reader that "she doesn't really look old." By the second sentence he is in the grip of his primary concern: to give the reader a certain impression of his mother. The third sentence is clearly in service of this goal.

Similarly, with the task force project, the writer wants to keep his description all on the surface. Here, he sticks to a superficial outline of the goal and the phases of the project—and assiduously avoids discussing what is really going on.

Now, when the writer turns from writing for someone else to writing for himself, the description itself radically changes. Instead of trying to give the reader a certain impression of his mother, he focuses on the living struggle of a person he loves. Instead of describing what an abstract task force was supposed to accomplish, he writes about what this one is actually doing.

Writing to discover = a problem solving tool

When we write to people, we usually do want to screen the information we give them. Yet, for ourselves, we also want a full understanding of the subject. We can use the second type of writing—writing to discover what we think and know about a subject—to get a deeper, broader view of it. *Then* we can choose what to say about it to others.

Writing to discover is a problem solving tool. It allows you to focus entirely on the subject, without being distracted by the impressions you want to give your reader.

How do you "write to discover"? It's simple. You just head the paper with your topic and then write down everything about it that comes into your head. Don't stop to rephrase, repunctuate, or reorganize your writing. Don't stop for anything at all. Just write. You're the only one who will see the result.

If your writing-to-discover leads you to some serious problems, as in the case of the task force above, continue to write about the topic until you can think of nothing further to say about it. Pause, read it over, and note in a few words the problems you unearthed, e.g. "staff resentment," "people lie about jobs," "no one trusts anyone else." Then, head a new sheet, "Solutions." Write down one of the listed problems and then focus all your energies on trying to come up with possible solutions to it. Do the same for the others. Use this as an inner brainstorming session; don't allow yourself to criticize the ideas you put down. Just state them and move on to others.

When do you write-to-discover? Whenever you want to see a situation as clearly and fully as possible—for instance, when you have to make a decision, report on a project, write a proposal, or prepare a talk. If you are using this writing to help you make a decision, it may be the only writing you need to do. If it's preparation for a report, proposal, or talk, it will be the raw material from which you can shape an ordered, readable, and appealing piece of communication. Whatever you choose to put into the final work will come from the truth as you know it, uncontaminated by any unconscious efforts to embellish or mask it for the reader's eye.

Writing to solve problems

Part 2: Rethink "persuasive writing"

Many problems have grown out of misguided attempts at "persuasive writing," mainly because writers have confused it with "fudge-writing."

By fudge-writing I mean exaggerating the merits of an idea or event and then putting blinkers on the reader so that he sees only that distorted picture. You put blinkers on your reader by:

- Omitting information
- Implicitly decrying all other points of view
- Suggesting that doing what you ask will benefit the reader enormously (even if it won't).

You may be thinking, "Oh, sure, I know that kind of writing. It makes up most of my 'junk mail' — sales letters trying to persuade me that I need to invest in a vacation home in the swamps of Dade County or that I should be flattered by the chance to pay $85 a year for an 'exclusive' credit card. I'm wise to that stuff. But I don't write it."

Well, think again. Almost all of us have written "that stuff" at one time or another, when we desperately wanted to make a certain impression or convince someone to do something for us. The technical memo that doesn't quite mention the fact that the project is a few days behind schedule. The letter requesting a paper for an upcoming conference that makes the conference sound a little more important and well-attended than it really is. The cover letter with a resume that makes the applicant's experience sound just a bit more relevant to the position than it really is.

Very few of us can say that we have never stretched a point to convince someone to do what we would like. But this so-called act of persuasion can lead to tremendous difficulties because it *hides problems* instead of solving them.

The report that skims over the few days' delay also skims over the reasons for the holdup and the new problems or schedule clashes that may grow out of it. These additional problems will not disappear just because we choose to ignore them.

The letter that slightly misrepresents the nature of the conference may lead to a serious problem with the person who agrees to give the paper. On arriving at the small, poorly attended meeting, this speaker may well end up resenting and mistrusting not only the letter-writer but also the organization he or she represents.

The cover letter that exaggerated the relevance of its writer's work experience may lead that person, once hired, to cover up his ignorance of certain crucial parts of the job and thereby cause problems and even danger.

Let's stop writing difficulties into our lives! Here are two ways to use writing to solve, not create problems:

1. *From now on, submit everything you write to a "problem check."* Search out problems that you have *not* mentioned and note them down for yourself. To do this, think back over the topic you're discussing, consider it step by step, and stop at the points you deliberately left out of your written discussion. Usually, these mark the areas where problems lurk. You may still decide not to mention them in your letter, memo, or report. But that doesn't mean *you* have to ignore them and avoid doing anything about them.

2. *Rethink persuasive writing.* Instead of glossing over a difficult fact, consider telling your reader both the really good news *and* the possible problems. No one expects a project to be totally problem-free—and no one really believes a report that suggests it to be so. If you begin your report by telling the reader the good things that have happened and then state clearly and concisely the problems that arose *and what you intend to do about them,* you'll write a much more believable, creditable, and truly persuasive report.

We have suggested to some companies that they include a "Challenges" section in each report. Challenges is a positive word for problems-to-be-solved. It implies that you are not afraid of mentioning or facing them and that you are preparing ways to meet and overcome them. The companies that have tried this addition have found that it works extraordinarily well *if* everyone does it and does it honestly.

The concept of accepting problems as positive challenges to be met rather than proofs of failure to be hidden is new to some businesses. Those firms that have made this new way of thinking part of their policy have found that it inculcates constant quality control, keeps problems smaller and under control, and, of course, improves communication and morale.

How about your company—or your department? Are you ready for the Challenge?

Writing to solve problems

Part 3: Writing and kaizen

In *KAIZEN: The Key to Japan's Competitive Success,* Masaaki Imai explains that kaizen (pronounced kai-ZEN) is a policy of ongoing, continuous improvement involving everyone in the company. He goes on to show, in example after example, how kaizen is indeed the major source of power that has propelled so many Japanese companies to a position far ahead of their Western counterparts.

There are four key elements of the kaizen approach:

1. **Kaizen is continuous and gradual, giving small but steady improvements**—as opposed to innovation, which aims at sudden, large improvements. In slow-growth economies or industries, kaizen may be the more effective mechanism for staying competitive.

2. **Kaizen is problem oriented.** It requires a corporate culture that welcomes problems as opportunities for improvement.

3. **Kaizen is process oriented—as opposed to results oriented.** It requires people to keep looking at *every* process for potential problems or shortcomings. The theory is that results will automatically improve if you improve all the processes.

4. **Kaizen is customer oriented.** It requires people to try never to pass any problem on to any customer and to make sure that problems that do arise will not happen again. The definition of "customer" here is: anybody who gets the results of your work, whether directly or indirectly. This includes people within your company.

As you can see, a fundamental part of kaizen is the acceptance—even welcoming—of problems as opportunities to improve. In this view, a problem is not an embarrassment to cover up. Instead, its discovery signals a great chance to make what you're doing even better.

If you're willing to look at problems as improvement opportunities—provisionally, if you like, to see where it leads—you can put kaizen to work for you through your writing.

Improvement opportunities

From now on, whenever you have a report to write, add a short section just for yourself and your co-workers. Call this section "Improvement Opportunities." It will consist of two parts, shown in outline form in the table.

You have been overseeing work on a new machine that the company bought on your recommendation to speed up production by 10%. The project was a great success; production rose between 18% and 20%. You are reporting on this success.

Successful projects are the ones most vulnerable to unnoticed problems. No one wants to think about the few things that didn't work out perfectly. The tendency is to figure that they'll sort themselves out. They rarely do. Instead, they either stay and develop or get passed on and transformed into different snags further down the line.

So—here's a sample "Improvement Opportunity" for your success report.

Review

What did we do? We used the new No. 3 machine to speed up production by 10%. Ron S., Al D., Carol N., and Carl P. worked on it. (Note what each one did, when, where, and why.)

What went well? We achieved 18-20% increase in four separate trials.

What went not so well? There were two minor accidents, put down to lack of familiarity with the machine.

What was my purpose? To achieve our goal of at least a 10% increase in production.

What were the needs of all the people involved? To achieve this goal *safely.*

How could I meet these needs better? By making safety as important as speed. I should look carefully at the accidents that took place and search for possible danger spots and ways to ensure the same accident doesn't occur again.

Refocus

New purpose. To achieve the same results with 100% safety.

An improvement idea. Do a whole trial purely to check safety at each phase.

Kaizen is a great import. If you bring it into your writing, in this or any other way, let me know. I'd love to hear how it works for you.

Improvement opportunities

Review
What did we do here?
Write a broad statement of what you did and why.
Go over the whole process minute, asking the "reporter's five questions" of each phase: Who was working on this project? What did each one do? When? Where? Why?
What went well?
What went not so well?
What was my purpose?
What were the needs of all the people involved?
How could I meet them better?
Refocus
Use the review to write:
A new purpose
An improvement idea for next time

Managing writing

Part 1: Market research

If you are a manager, you're held accountable for the quality and timeliness of the written documents issued by your group. It may seem unfair, given how much people's writing ability is out of your control—but it's a fact of corporate life.

The price of haphazard management

Unfortunately, many managers seem to handle this area of their work haphazardly. They never get out of the firefighting mode. I wonder if you are experiencing some of the firefighting syndromes yourself: Are you continuously rewriting half the reports in your group? Are you always scrambling to beat deadlines because your people started their reports too late? Do you often take beatings for politically insensitive reports? Do you keep on correcting the same mistakes by the same employees? Do you alternatingly plead, command, and threaten—all with no result other than rebellious or demotivated employees?

But you *can* get out of that cycle of wasted time and emotion. What's more, the steps are the same you'd use for a technical project—so you're already thoroughly familiar with them:

1. Evaluate the quality of writing in your group.
2. Make a detailed improvement plan.
3. Implement the plan.

In this series of columns, we'll look at those three steps in detail.

Evaluating your group's quality of writing

First you need to assess the problem. This has two aspects. One, you need to know how satisfied your regular *readers* are with the documents produced by your group. Two, you have to look at each person in the group and get an objective picture of writing weaknesses to overcome.

The first part is really quite easy. The main problem is that most managers never even think of doing this. They just assume they know what kinds of documents and correspondence their customers want. The result in most technical organizations: guidelines on format, style, and substance that miss the readers' needs by a mile. Yet your readers' satisfaction obviously should be the ultimate criterion in assessing the quality of writing. Just as you wouldn't dream of launching a product without doing market research, you shouldn't grind out reports without doing basic research on your readers.

A simple market survey

I'm not saying you need a complicated formal "survey" with lengthy questionnaires that make people yawn and groan. Just pick two or three clients whom you judge to be cooperative and thoughtful. Then devise your own questionnaire to guide the conversation. Here are some of the questions you'll probably want to ask:

- Are our reports timely and frequent enough?
- How quickly and easily can they be read?
- Is all the important information generally there?
- How much irrelevant information is there?
- How good is the *attitude* that comes across?
- Do the reports inspire *confidence* in the writer's work?

It's also a good idea to include an open-ended question that lets the client air general feelings—say: "What are your biggest concerns in this whole area of written communications?" Some of your most valuable information may come out of such questions.

Finally, lift the whole exercise out of the abstract realm by briefly going over a real report—preferably one that you weren't very happy with yourself. Ask the client to tell you just how he "digested" that report: where he started, how he went about finding the information that really mattered to him, what parts bothered him, which points were unclear, how he felt about the project as a result of the report, and so on. You'll be surprised to what extent the abstract information you received will be modified in this phase. People often don't realize what their real problems and concerns are until they're forced to walk through a concrete example.

The whole conversation usually won't take more than half an hour. But it may give you some surprising insights and change many of your ideas on format, style, and frequency of reports.

There is another important benefit of this simple market research: you can now go to your people and say: "This is what our customers say about our writing. This is what they want and need. And this is what they *don't* want." It's no longer you making arbitrary demands—it's a matter of you and your people working together to satisfy your customers. That's an important first step toward eliminating confrontation and resentments and moving toward a long-term improvement plan.

Next month, we'll look at the second aspect of step 1: analyzing individual writing problems in your group. By that time, I hope, you'll have finished your own market survey.□

Managing writing

Part 2: Evaluating individual writing problems

Last month we started to look at a systematic way to improve written communication in your group. This involves three steps:

1. Evaluate the quality of writing in your group.
2. Make a detailed improvement plan.
3. Implement the plan.

The first step, evaluation, has two aspects. First, find out how your **customers**—those who regularly read your group's documents—view the quality of writing. (This part, which we covered last month, is your "market research;" it establishes broad goals for your improvement effort.) Second, evaluate the strengths and weaknesses of each writer in your group. This is our topic this month.

Writers and their problems

People can have lots of problems with writing. Most problems with writing fall under three general categories:

1. Writing skills
2. Attitude
3. Knowledge or information.

Each category calls for different remedies.

Your best approach is to start a worksheet for each writer in your group. Your worksheet should have three columns, titled "Problem Areas," "Assessment," and "Improvement Plan." In the left column, under "Problem Areas," list the following items:

- Writing skills—Distilling essence, organization, style/tone, grammar and punctuation, and spelling
- Attitude—Hate/fear of writing, carelessness, no appreciation of the importance of high-quality communication, poor time management, demotivated because nobody responds to memos/reports, demotivated because of arbitrary editing or criticism, frustrated about lack of time for writing (poor planning by managers)
- Knowledge/information—No understanding of clients and their needs, not aware of group's standards for documents, not aware of political subtleties/ramifications of a project or communication.

As I mentioned, each class of problems tends to call for different remedies. Generally, *skill problems* call for skill training; *attitude problems*, for counseling, motivational training, and changes in managers' behavior; and *knowledge/information problems*, for clear standards and guidelines, regular sharing of important information, and more direct contact between customers and your people. But more on this next month.

Getting an objective evaluation

You could of course sit down and fill out the whole worksheet for each person, then move right on to remedial action. This, however, would probably not give you an objective base. I suggest that you take a whole month or even longer to build an objective record on each writer.

During that period, don't just edit, criticize, and discuss. Make a written note on exactly what you discussed. Keep a copy of your editorial changes. And look at the *reason* for each editorial change (grammar, style, organization, content, etc.).

Also, during this period, *talk* to your people to find out more about their attitudes, and the sources of those attitudes.

You may find that there are some areas you just cannot evaluate objectively. For instance, your own writing and editing skills may not be strong enough to identify people's style and grammar errors. In that case, I'd suggest enlisting the help of a star writer in your group or, if there is no such person, a competent communications consultant.

A look at yourself

As you continue this process of evaluation, you should also fill out a worksheet for yourself. It should contain all the items in the general worksheet, plus the following points:

- How good are your counseling skills? Do you listen well? Do people trust you? Do they listen to you?
- Are you budgeting realistic amounts of time for people to write?
- Do you prepare people well for a writing job by sharing all important information, including political issues?
- Are your editorial changes thoughtful and necessary?

As the assessments on your worksheets begin to solidify, you'll be better prepared to develop a really strong, targeted improvement plan. You'll have detailed records to back up your judgment—a definite plus when it comes to counseling or other remedial steps.☐

Managing writing

Part 3: Planning for improvement

In the last two months, you saw how to evaluate the quality of writing in your group by interviewing some of your main readers and then assessing the strengths and weaknesses of each writer. If you started your program with the first article of this series, you should have a worksheet for each writer, including yourself. Have you filled in the two left columns, titled Problem Areas and Assessment? Then you're ready to move on to column 3, Improvement Plan.

A three-part plan

A good, specific plan is essential—it will make the difference between wasted training and an effective program with lasting results. Your plan will need three major elements: (a) what to do for each writer, (b) what to do for yourself, and (c) what to do for the group.

Individual plans. Many managers can see the need for group training in writing, but few think of starting with an individual plan for each person. Yet this is the key to success—it gives a rational basis for both the group plan and your personal plan, and it helps you monitor progress in a specific, meaningful way. **Table 1** lists some possible remedies for writers' problems.

Your own plan. Do certain management-related items come up repeatedly in the individual plans? Then they should be the central elements of your personal plan. You may have to improve your resource planning (budget more writing time for subordinates), become more willing to share important information, or curb the urge to rewrite everything. You may even have to get some help with your counseling, listening, and communication skills. If there are no clear guidelines and standards for documents, you will have to attend to that, too—though you may delegate the job to a task force (a good idea, for many reasons).

Finally, if your own writing and editing skills are weak, you may want to join the training program you're choosing for your people. Far from damaging your authority, this will earn you tremendous respect and do a lot toward improving trust and communication.

The group plan. This is simply a consolidation of the individual plans. Do most of your people need skills training and some counseling plus motivational training? Then a custom-designed writing/consulting program for the whole group may be the most effective solution. Or do many people in the group lack a clear understanding of customers' needs and constraints? Then a policy of encouraging direct customer contact and giving people responsibility for entire projects will help.

Your resources

Achieving your goals doesn't necessarily mean calling in training consultants. Don't overlook internal resources. For example, can the Manager of Budgets and Controls give some workshops on how to put together a superb program proposal? Can a task force of competent writers (perhaps drawn from several groups) develop clear standards for reports and technical memos and present those standards in a series of seminars? Or are there some other qualified people in your organization who would be glad to get involved in your improvement program?

Look at all the resources available to you, then decide on the best mix. Next month, we'll discuss some guidelines on using those resources, including outside trainers.□

1. Writers' problems and possible remedies

Problem Areas	*Possible Remedies*
WRITING SKILLS (substance, organization, style, grammar, punctuation, spelling)	Skills training Self-help groups (mutual editorial support)
ATTITUDE	
Hates/fears writing even though not a bad writer	Assign special writing projects with high status to give positive experience
Doesn't care about the quality of communication	Counsel; motivational training; expose more to customers; set an example
Demotivated because of no response to writing	Give more feedback on writing; find ways to get more responses from customers
Poor time management (procrastinates)	Ask to submit typed notes of key project discussions by next day
Frustrated about lack of time for writing	Budget more time for writing
Demotivated because of arbitrary editing or criticism	Minimize editing; base all editing/criticism on *objective* standards
KNOWLEDGE/INFORMATION	
Doesn't understand clients and their needs	More direct customer contact; more responsibility for seeing whole projects through
Not aware of political implications of work/communication	Share political and other information regularly, at beginning of a project
Not aware of group's standards for documents	Train to follow standards; develop or update standards/guidelines if necessary

Managing writing

Part 4: Implementing your improvement plan

You've clarified the requirements of your main readers, evaluated the quality of writing in your group, and worked out a sensible plan for improving things. Now it's time to put the plan into action—starting with yourself.

Implementing your personal plan

There are three major things you can do to contribute directly to the improvement process:

1. Giving people more feedback on memos and reports.
2. Assigning special projects that involve writing can be helpful where morale and motivation are a problem.
3. Giving people more direct customer contact and responsibility for entire projects will improve their understanding of clients' needs.

Here are some other elements that may need attention:

- **Sharing information.** This means giving up some of your power. The payback is not only better writing but better work, not to mention morale. You can make information sharing a *routine* with a "Project Start-up Sheet." One item would be "Information to share making the project successful."
- **Coaching.** If you're really skilled at both editing and coaching, you may be able to help people directly. But *cooperation*, rather than criticism, is the key word here. You might say, "John, let's work on this report together." Instead of spending a lonely half-hour editing John's report, you may sit for 40 minutes together—and accomplish a lot of very useful things besides improving that report.

Be realistic, though. If your coaching skills are weak, get some training.

- **Dealing with procrastinators.** When you first discuss a project with a procrastinator, suggest firmly that he take notes. At the end of the session, review the notes and firm them up—especially the formulation of objectives. Demand to see a rough typed version by the next day. This way, the procrastinator has started writing on day 1!

Using internal resources

As mentioned last month, try to tap star writers from other groups. Often, all that's needed is a compliment. They may be so flattered by your appreciation that they give up great chunks of their valuable time—all with a big smile.

Self-help groups are another useful approach. People can help each other edit. It's easier to be objective with other people's pieces. However, such group efforts are usually more effective when people get some training in how to edit.

Working with outside trainers

What should you look for when bringing in outside trainers? Above all, commit them to *results*. The results you want are the things you put down as goals in your plan. The trainers have to understand those goals, and they must *custom-design* the training so it achieves those specific goals. Stay away from "one-shot" training—you can't achieve your goals unless the program goes on for some time. Finally, the training should not just consist of workshops but include *individual work* with the trainees. People need lots of competent, nonthreatening feedback if they are to make real progress.

Developing standards and guidelines on writing

It's best to assign this to a task force. Why? Because it will be a tremendous learning experience. Don't hand this task just to the competent writers; let some weaker ones participate so they can start feeling involved.

There is no question that you need clear standards and guidelines. If you don't have competent writers who can develop them for you, and you're not skilled in this area either, then you'll have to consider turning this over to some outside consultants or trainers.

Monitoring progress

You'd be amazed how many managers spend lots of time and money on training—never following up on the results. Why? Because they're afraid to find out that they made a poor decision. They'd rather look the other way and hope for the best. This may be great for the trainers—but not for the company, which foots the bill. So put follow-up dates on your calendar *now*—and stay in touch with the trainers, the "students," and the supervisors.

Here are some of the things you'll want to reevaluate: Are supervisors editing less? Is there less friction because editing is less arbitrary? Are readers more pleased with the final product? Do they get reports more quickly? Are they saving time by having to read less material that is irrelevant to them? Do people feel better about writing? Do they write more quickly? And are they getting more out of it—i.e., does it contribute more to their primary work?

If you followed the systematic approach we discussed in this series, I think you'll be pleased with the answers to all these questions. Good luck—and let me hear from you.□

Principles from poetry

Part 1: Persuasion

It's the season when everything feels different—the time, the air, the town, the people, even we, ourselves. We think differently at this time of year. Instead of planning meetings, we're trimming trees. Instead of wondering how to save money, we're out spending it on family and friends. Instead of worrying—we're hoping, a little.

In keeping with the season of change, I'm going to suggest that you begin a change in reading habits. To your regular diet of technical or business material, add a little poetry. Wait, please—don't stop reading this yet! I'm not suggesting this only to offer you the aesthetic and spiritual gifts of poetry. Poetry will help you write better memos, letters, and reports.

Great poetry releases the power in ordinary words and makes them resonate. The poets take all the principles of writing—persuasion, clarity, organization, force—and exploit them to the maximum. In a few words, they can tell the story of the world. To discover the possibilities in language and use it to transmit your message with real clarity and power—you must read poetry.

Poetry and persuasion

To his dying father, Dylan Thomas wrote:

> Do not go gentle into that good night.
> Rage, rage against the dying of the light.

Though the message is complex and perhaps disturbing, the words themselves are so simple and familiar to us that their images flash up instantly before our eyes; we can't avoid them.

In the first line, the words "go", "gentle," and "good night" (a good night, or good night as meaning good-bye), give an image of peace, quiet, harmony (through the repetition of the initial sound), and safety. The whole line is lulling, soothing, like a siren's song. It represents death as the victim would like to see it: easy, cosy, quiet.

But the words of the second line—"rage" and "dying of the light"—project an astonishing contrast of violence and emptiness. The second line takes the images created by the first and smashes them to pieces. This is not a *good night*—it's *the dying of the light*. The spirit of a man must fight against it to remain human.

Now, what does all this have to do with the writing you have to produce every day? These two lines of Dylan Thomas' poem embody a fundamental principle of persuasion: First present your opponent's argument in the way he sees it; then knock it down with the force of your own.

In the first line, Dylan Thomas presents death as the dying man might view it: a welcoming "good night." In the second, he sets up a much more powerful image of death as something against which we must fight. The second image conquers the first by its sheer force.

Using the principles of persuasion

Suppose you want to convince your company to send you and some others on a fact-finding trip abroad. Let's imagine that you write or say simply, "This trip could be of benefit to our work in drying, because we would be able to observe the new machines in the Munich and Gottenberg labs firsthand." Now, suppose your executives respond that they know that, but this is a year for cost-cutting. You've already stated your argument. Now all you can do is repeat it. That's not effective persuasion.

But if you start by presenting their arguments, you can then go on to knock them down with the force of your own! And your opponents will be left with little force on their side. For example, you could start by writing, "We are all cutting costs this year, in an effort to increase the company's strength and productivity." As you elaborate on this statement, you present your opponent's objections right from the start. They won't be able to surprise you with them later.

Then you demolish their objections with the greater force of your own message:

> To forge ahead of our competitors, especially in these difficult times, we *must* have the best, most efficient, and most productive papermaking methods available. We cannot afford to lag behind. The XYZ Company in Germany has invited a team of our researchers to observe their new PQR dryers, which promise to raise production substantially. I believe we must observe these machines firsthand, and soon.

Now, over the next year, you may remember parts of this column. Perhaps you'll recall the principles of persuasion it enunciated. But—if you read Dylan Thomas' poem, "Do Not Go Gentle Into That Good Night," a few times, you'll never forget those electric lines. And they'll show you how persuasion works, in just 16 words.

Principles from poetry

Part 2: The power of the unexpected

Robert Frost wrote:

> Something there is that doesn't love a wall.

The line resonates in our memories. It does *the unexpected*; it turns the normal order of words around. But if Frost had written simply, *There is something that doesn't love a wall*, his words would have lost their power to arrest our attention and make us think about their meaning.

T. S. Eliot wrote:

> Let us go then, you and I,
> When the evening is spread out against the sky
> Like a patient etherized upon a table;

The first two lines are lilting and easy. They even rhyme. They suggest a dreamy evening walk in the glow of twilight. We expect a color-filled, romantic description of the sunset. So we jump when we get to the third line. Eliot's etherized patient is the opposite of the image we were expecting, and quite unlike any we have ever seen about the evening. It makes us stop in our reading and think, "Why is he throwing this at me? What is he talking about?"

Shakespeare wrote:

> That time of year thou mayst in me behold
> When yellow leaves, or none, or few, do hang
> Upon those boughs which shake against the cold,
> Bare ruined choirs, where late the sweet birds sang.

He starts with a fairly conventional comparison of old age and winter. We focus on the *visual* connection between a winter tree touched with a few yellow leaves and an aged head with a few thin hairs.

And then Shakespeare hits us with his lightning-line:

> Bare ruined choirs where late the sweet birds sang.

And we stop, stunned by the power of his words. The image is totally unexpected. Shakespeare has switched from present sight to remembered sound.

With this line the rhythm of the poem suddenly changes. Each of the first three lines followed a regular iambic pentameter: (-/-/-/-/-/). This riveting fourth line stops the easy trot cold. All but two words—*where* and *the*—are heavily accented:

> Báre ruíned choírs where late the sweet bírds sang.

It is impossible not to pause at this marvelous, unexpected line and let its sounds and meaning echo in your mind.

Using the unexpected in business writing

These poems point to a great tool for all writers: surprise, or the unexpected. To get your reader to pay attention to a particular point, give him the unexpected.

As you saw in the three poems, the poets all did something the reader did not expect them to do, and so captured his attention. They wielded the unexpected in three ways. Frost used *unexpected word order*. Eliot made an *unexpected statement*. And Shakespeare produced an *unexpected image* and *unexpected rhythm*.

Let's suppose you are submitting a proposal to your boss. You want to engage some outside trainers for your programmers. Until now, one person in the company has been training them, and he has neither the time nor the expertise to do it properly.

Now, you want to focus your reader's attention on the fact that the programmers are not making full use of the computer because they don't know how. This is a perfect opportunity to use the unexpected.

You might follow Robert Frost and use unexpected word order:

> 35% of capacity. That's not enough use for a $2,000,000.00 computer system.

Or, you could make an unexpected statement, a la T. S. Eliot:

> We don't need our computer anymore!

Once you've shocked them into attention, you can elaborate:

> Since our programmers are using it to only 35% of its capacity, we might as well trade it in. Of course, there is an alternative. We could train our programmers to use *100%* of the computer.

Although you may rarely have the opportunity to create an unexpected image, as Shakespeare did, you can often use his technique of *unexpected rhythm*. Suppose you start:

> At present, our programmers are operating the $2,000,000.00 computer system at only 35% of its capacity. We are wasting a valuable asset, because our people are insufficiently trained to take full advantage of it.

You can then hit the reader with your demand:

> They need more skills. And more time to learn.

This series of monosyllabic words contrasts heavily with the two- or three-syllable words in the preceding paragraph. The last two sentences hit the reader like a thud of drumbeats after a tune of violins. He has to pay attention.

Exploit the power of the unexpected! It's another great principle of writing and you can learn from poetry.

I hope you are reading some poetry now, for pleasure as well as principles. If you discover any principles from poetry that you'd like to share, please send them to me. I'll publish them, with your name, in an upcoming issue.

Poems quoted: "Mending Wall," Robert Frost; "The Love Song of J. Alfred Prufrock," T.S. Eliot; "Sonnet 73," William Shakespeare.

Principles from poetry

Part 3: Visualization

Do you use poetry in your daily life?

If you've been following this series, I hope you have been reading a little poetry once in a while, instead of the latest disaster in the newspaper—even, just occasionally, instead of the stock reports or the sports page! But whether or not you consciously let poetry into your life, you almost certainly use it or hear it *unconsciously.*

If you call someone *a snake in the grass,* you are citing Virgil: "There is a snake hidden in the grass."-Eclogues II.

If you remark that a colleague has a spotless reputation, you are talking Shakespeare: "The purest treasure mortal times afford/Is spotless reputation."-King Richard II.

If a friend calls you a tower of strength, he is calling on Tennyson: "O fall'n at length, that tower of strength/ Which stood four-square to all the winds that blew."-Ode on the Death of the Duke of Wellington.

A lurch into *the jaws of death* also draws on Tennyson, while the act of *gilding the lily* refers to Shakespeare. Hundreds of cliches were originally lines of poetry. Why do we keep using them? I believe because they have the power of *visualization.*

Visual language is powerful language

Visual language transmits a message by showing the reader or listener a picture. It has tremendous impact. You can ignore a dry statement but not an emotional image. That's one reason poetry is so powerful; it presents images that shatter our complacency and force us to pay attention.

Look at all those expressions above. Each is an expression of vibrant imagery. You can *see* a snake skulking in high grass . . . a terrible monster with open jaws . . . a lily tipped with gold.

We try to profit from this power when we run to the old cliches. But they don't work any more. The images that originally stunned readers into observance have now become so commonplace that we barely see them when we use or hear the expressions.

No, to use visualization effectively, you have to start thinking of fresh images. Think visually. Try to see an image that transmits your thought. Then use it.

For instance, if I remarked, "I spend too much time gossiping at trivial social occasions," you would probably nod sympathetically and then forget it. But if I were T. S. Eliot and said, "I have measured out my life in coffee spoons . . ." you'd pay attention. You would see the image depicted and react to it.

Visualization at work

Suppose a colleague has delayed in sending out an important document. You are writing to him to ask him to get onto it—fast. If you're not thinking visually, you might write something like: "I urgently need the document. Please send it to me now." These statements transmit your message, but not powerfully.

However, suppose you start thinking about the situation visually. Close your eyes and see how you've been reacting to this delay. Then show your colleague that image:

> "Every day, at 11 a.m. and again at 2 p.m., I sit at my desk waiting for the mail cart. Each time I hear it roll down the office halls, I think, 'This time I'll get the document from Joe.' Ritchie puts the mail on my desk, and I go through it eagerly, looking for an envelope with your return address on it. But—it's never there.
>
> Please, Joe, send it to me now. I can't take this twice-a-day trauma much longer!"

Joe will see you at that desk, eagerly awaiting the document he has been too lazy to send—and then looking dejectedly at the mail before you. With that image before him, he'll probably be moved to get that document off to you right away.

Or, let's say you are a sales manager. One of your representatives has been late twice for a meeting with an important prospect. To get him to shape up, you want to tell him that the loss of this sale will hurt his career at your firm. If you don't think visually, you may say: "If you spoil that sale for us, Eric, your chances for advancement around here are going to be pretty small." Eric will get the message, but not in the solar plexus. But now try thinking visually. How can you make Eric see what will happen if he doesn't salvage this sale? "If they slam the door in your face over there, Eric, you'll come back to a whole row of closed doors back here, too." Eric will see himself being shut out of the potential client's offices—and then out of the offices he had hoped to occupy at home. He'll react.

When you need especially powerful language, try using visualization. Read poetry, especially Shakespeare, William Blake, and Robert Frost. You'll see the images shining off the pages. Identify the abstract thought behind the image. Then notice how the poet conveyed it, visually.

Begin doing some visual thinking yourself. See yourself reacting to something. Then describe this reaction in some detail, so that other people can see it, too. Instead of outlining the consequences of an action, present a scene that shows what will happen.

You'll find people start listening to you more closely—and paying much more attention to what you say.

Principles from poetry

Part 4: Creative thinking

This fourth and final column of "Principles from poetry" will show you how poetry can help you think creatively. If you use the technique shown here and read a little poetry regularly, you will start thinking differently. You'll go beyond limiting conclusions. You'll feel as if you are seeing around corners.

And your writing will become brighter and more arresting because it will reflect your new thoughts. Remember, creative writing grows out of creative thinking. The pen simply turns the mind's ideas into words.

The three-part technique for creative thinking that poetry can help us develop is this: (a) See the details. (b) Look at things differently. (c) Make a new connection.

See the details

To get new, useful ideas about a product or a subject, you must first see it in all its parts. By noting some previously ignored details, you may well discover a successful approach to a problem.

Many poets celebrate the details of things and alert you to see them yourself. For this skill, read particularly the poetry of Robert Frost and any poems in the Japanese *haiku* tradition (including *haiku* poems by Ezra Pound).

Spring Breeze
These morning airs -
one can see them stirring
caterpillar hairs!
-Buson

Ts'ai Chi'h
The petals fall in the fountain,
the orange-colored rose-leaves,
Their ochre clings to the stone.
-Ezra Pound

Look at things differently

Now, reject the conventional, "obvious" view of the subject. Look at it upside down or sideways or the other way round.

Poets show you how to see things differently. For instance, the normal view of the seasons is that winter is cruel and spring, kind. But T. S. Eliot opens "The Waste Land" saying: *April is the cruellest month. . . .*

And Shakespeare turns the conventional comparisons of "romantic" poetry upside down in Sonnet 130: *My mistress' eyes are nothing like the sun; Coral is far more red than her lips' red . . . .*

Make a new connection

Finally, make a connection between this new view of your subject and a truth or a possible answer.

If Eliot had simply announced that April was the cruellest month and left it at that, we might have dismissed his statement as idiosyncratic at best. But he goes on to make a connection between this odd statement and the truth that memories of happy days hurt badly when those days are gone, forever:

April is the cruellest month, breeding
Lilacs out of the dead land, mixing
Memory and desire, stirring
Dull roots with spring rain.

And Shakespeare connects his unconventional thought to the truth that love makes the loved one more wonderful than any of the earth's beauties:

And yet, by heaven, I think my love as rare
As any she belied with false compare.

Thus, both poets make sense of their seemingly odd view.

Creative thinking in the paper industry

Hardwood trees were once considered the "weeds" in paper forests, since their fibers were too short to make strong paper.

But one day, someone applied creative thinking to hardwoods. The creative thinker (whom we'll dub "C.T.") *saw the details* of the trees, and noticed that the fibers were not only short and weak but also extremely fine.

Next, C.T. *looked at the hardwoods differently.* Upturning the conventional conclusion that short, weak fibers make "bad" paper, C.T. said boldly, "These fine fibers will make good paper!"

Then C.T. *made a new connection:* "These fibers can fill up the spaces between the longer, thicker softwood fibers. They'll greatly improve the smoothness and printability of the paper!" And hardwoods weren't weeds any more.

Or, consider the evolution of sack paper. Conventional wisdom said, "The stronger the paper, the better it is for carrying cement." But then C.T. came along.

C.T. *saw the details* of cement sacks; the strongest ones were still cracking in places under the heavy load.

C.T. then looked at the sacks differently. Others said, "We need stronger paper to contain the load." C.T. said, "No, we need more stretchable paper to give with the load."

Then C.T. *made the new connection:* "Strong paper doesn't give; that's why the sacks are cracking. To get more stretchable paper—we must make it not stronger, but more stretchable. Even if we *lose* some strength."

And Clupac was born.

Creative thinking produces creative writing, inventive solutions, and wise business decisions. You can start thinking more creatively now. Try to apply the three-step technique of creative thinking to your work. And read some poetry. I hope I have convinced you to make it part of the language of business!

Afterword

I hope these columns have brought you some useful ideas and a sense of the possible power and beauty of the language we use every day. If we don't try hard to communicate honestly and completely with one another, we may indeed raise a new Tower of Techno-Babel. But if we do try, we may at last achieve understanding among ourselves and real quality in our work.

Thank you for caring about the language of business.

Conversion Factors for SI Units

Quantity or Test	Value In Trade or Customary Unit X	Conversion Factor =	Value in SI Unit	Symbol
Area	square inches	6.45	square centimeters	cm^2
	square feet	0.0929	square meter	m^2
	square yards	0.836	square meter	m^2
	acres	0.405	hectares	ha
Basis Weight * or Substance (500-sheet ream) or Grammage* when expressed in g/m^2	lb (17x22-500)	3.760	grams per square meter	g/m^2
	lb (24x36-500)	1.627	grams per square meter	g/m^2
	lb (25x38-500)	1.480	grams per square meter	g/m^2
	lb (25x40-500)	1.406	grams per square meter	g/m^2
	pounds per 1000 sq ft (Paperboard)	4.882	grams per square meter	g/m^2
Breaking Length	meters	0.001	kilometers	km
Burst Index	$\frac{g/cm^2}{g/m^2}$	0.0981	$\frac{kPa}{g/m^2}$	
Bursting Strength	pounds per square inch	6.89	kilopascals	kPa
Caliper	mils	0.0254	millimeters	mm
Concora Crush	pounds	4.45	newtons	N
Edge Crush	pounds per inch	0.175	kilonewtons per meter	kN/m
Energy	British thermal units (Btu.)	1055	joules	J
Flat Crush	pounds per square inch	6.89	kilopascals	kPa
Force	kilograms	9.81	newtons	N
	pounds	4.45	newtons	N
Length	angstroms	0.1	nanometers	nm
	microns	1	micrometers	um
	mils	0.0254	millimeters	mm
	feet	0.305	meters	m
Mass	tons (2000 lbs)	0.907	metric tons	t
	pounds	0.454	kilograms	kg
	ounces (avd p)	28.3	grams	g
Mass per Unit Volume	ounces per gallon	7.49	kilograms per cubic meter	kg/m^3
	pounds per cubic foot	1.60	kilograms per cubic meter	kg/m^3
Puncture Resistance	foot pounds	1.36	joules	J
Ring Crush	pounds (for a 6 in. length)	0.0292	kilonewtons per meter	kN
Stiffness (Taber)	gram centimeters (Taber Units)	0.0981	millinewton meters	mN•m
Tear Strength	grams	9.81	millinewtons	mN
Tensile Breaking Load	pounds per inch	0.175	kilonewtons per meter	kN/m
	kilograms per 15 millimeters	0.654	Kilonewtons per meter	kN/m
Volume, Fluid	ounces (US Fluid)	29.6	milliliters	mL
	gallons	3.79	liters	L
Volume, Solid	cubic inches	16.4	cubic centimeters	cm^3
	cubic feet	0.0283	cubic meters	m^3
	cubic yards	0.765	cubic meters	m^3

*See TAPPI Technical Information Sheet 0800-01.

TAPPI Vision Statement

What is TAPPI?

Founded in 1915, TAPPI is the world's largest professional society of executives, operating managers, engineers, scientists, and technologists serving the paper and related industries. Total membership is approximately 32,000 with some 80% residing in the United States. The remainder live in 76 other countries.

TAPPI is renowned for its industry publications. Members produce technical books, reports, conference proceedings, course notes, home study courses, and videotapes through TAPPI PRESS. *Tappi Journal,* distributed monthly to all members, is the leading publication for technical information on the manufacture and use of pulp, paper, packaging, and converted products. Through TAPPI, Association members develop, update, and publish test methods and technical information sheets on which much of the industry depends to analyze its products and processes.

TAPPI sponsors a variety of technical conferences, seminars, and short courses to foster worldwide technical information exchange and enhance the professional development of members.

For membership information, to order any of TAPPI's professional development products, or to register for a meeting, please call TAPPI's toll-free Service Line: 1-800-332-8686 (U.S.); 1-800-446-9431 (Canada).

TAPPI's Vision

We are a global community of motivated individuals who lead the technical advancement of the paper and related industries.

Together...

- We provide outstanding educational and professional growth opportunities.

- We serve as a worldwide forum to exchange technical information, promote research, and recognize individual achievement.

- We create success by the quality, timeliness and innovativeness of our products and services.

Integrity and fellowship characterize our association.